Firefighter Fatalities in the United States in 2009

Prepared by
U.S. Department of Homeland Security
Federal Emergency Management Agency
U.S. Fire Administration
National Fire Data Center
and

The National Fallen Firefighters Foundation
www.firehero.org

In memory of all firefighters

who answered their last call in 2009

To their families and friends

To their service and sacrifice

TABLE OF CONTENTS

ACKNOWLEDGMENTS

This study of firefighter fatalities would not have been possible without the cooperation and assistance of many members of the fire service across the United States. Members of individual fire departments, chief fire officers, wildland fire service organizations such as the U.S. Forest Service (USFS), the National Park Service (NPS), the Bureau of Land Management (BLM), the Bureau of Indian Affairs (BIA), the U.S. Fish and Wildlife Service (FWS), as well as the U.S. Department of Justice (DOJ), the National Fire Protection Association (NFPA), and many others contributed important information for this report.

The National Fallen Firefighters Foundation (NFFF) was responsible for compilation of a large portion of the data used in this report and the incident narrative summaries found in the Appendix.

The ultimate objective of this effort is to reduce the number of firefighter deaths through an increased awareness and understanding of their causes and how they can be prevented. Firefighting, rescue, and other types of emergency operations are essential activities in an inherently dangerous profession, and unfortunate tragedies do occur. This is the risk all firefighters accept every time they respond to an emergency incident. However, the risk can be greatly reduced through efforts to improve training, emergency scene operations, and firefighter health and safety initiatives.

BACKGROUND

For 33 years, the U.S. Fire Administration (USFA) has tracked the number of firefighter fatalities and conducted an annual analysis. Through the collection of information on the causes of firefighter deaths, the USFA is able to focus on specific problems and direct efforts toward finding solutions to reduce the number of firefighter fatalities in the future. This information is also used to measure the effectiveness of current programs directed toward firefighter health and safety.

Several programs have been funded by USFA in response to this annual report. For example, USFA has sponsored significant work in the areas of general emergency vehicle operations safety, fire department tanker/tender operations safety, firefighter incident scene rehabilitation, and roadside incident safety. The data developed for this report are also widely used in other firefighter fatality prevention efforts.

One of USFA's main program goals is a 25-percent reduction in firefighter fatalities in five years and a 50-percent reduction within 10 years. The emphasis

placed on these goals by USFA is underscored by the fact that these goals represent one of the five major objectives that guide the actions of USFA.

In addition to the analysis, USFA, working in partnership with the NFFF, develops a list of all onduty firefighter fatalities and associated documentation each year. If certain criteria are met, the fallen firefighter's next of kin, as well as members of the individual's fire department, are invited to the annual National Fallen Firefighters Memorial Weekend Service. The service is held at the National Emergency Training Center (NETC) in Emmitsburg, MD, during Fire Prevention Week in October of each year. Additional information regarding the Memorial Service can be found at www.firehero.org or by calling the NFFF at (301) 447-1365.

Other resources and information regarding firefighter fatalities, including current fatality notices, the National Fallen Firefighters Memorial database, and links to the Public Safety Officers' Benefit (PSOB) Program can be found at www.usfa.dhs.gov/fireservice/fatalities/

This report continues a series of annual studies by USFA of onduty firefighter fatalities in the United States.

The specific objective of this study is to identify all onduty firefighter fatalities that occurred in the United States and its protectorates in 2009 and to analyze the circumstances surrounding each occurrence. The study is intended to help identify approaches that could reduce the number of firefighter deaths in future years.

Who is a Firefighter?

For the purpose of this study, the term firefighter covers all members of organized fire departments with assigned fire suppression duties in all 50 States, the District of Columbia, and the Territories of Puerto Rico, the Virgin Islands, American Samoa, the Commonwealth of the Northern Mariana Islands, and Guam. It includes career and volunteer firefighters; full-time public safety officers acting as firefighters; fire police; State, territory, and Federal government fire service personnel, including wildland firefighters; and privately employed firefighters, including employees of contract fire departments and trained members of industrial fire brigades, whether full or part time. It also includes contract personnel working as firefighters or assigned to work in direct support of fire service organizations (i.e., air-tanker crews).

Under this definition, the study includes not only local and municipal firefighters, but also seasonal and full-time employees of the USFS, NPS, BLM, BIA, FWS, and State wildland agencies. The definition also includes prison inmates serving on firefighting crews; firefighters employed by other governmental agencies, such as the U.S. Department of Energy (DOE); military personnel performing assigned fire suppression activities; and civilian firefighters working at military installations.

What Constitutes an Onduty Fatality?

Onduty fatalities include any injury or illness sustained while on duty that proves fatal. The term "onduty" refers to being involved in operations at the scene of an emergency, whether it is a fire or nonfire incident; responding to or returning from an incident; performing other officially assigned duties such as training, maintenance, public education, inspection, investigations, court testimony, and fundraising; and being on-call, under orders, or on standby duty except at the individual's home or place of business. An individual who experiences a heart attack or other fatal injury at home while he or she prepares to respond to an emergency is considered on duty when the response begins. A firefighter that becomes ill while performing fire department duties and suffers a heart attack shortly after arriving home or at another location may be considered on duty since the inception of the heart attack occurred while the firefighter was on duty.

On December 15, 2003, the President of the United States signed into law the Hometown Heroes Survivors Benefit Act of 2003. After being signed by the President, the Act became Public Law 108-182. The law presumes that a heart attack or stroke are in the line of duty if the firefighter was engaged in nonroutine stressful or strenuous physical activity while on duty and the firefighter becomes ill while on duty or within 24 hours after engaging in such activity. The full text of the law is available at http://frwebgate.access.gpo.gov/cgi-bin/getdoc.cgi?dbname=108_cong_public_laws&docid=f:publ182.108.pdf

The inclusion criteria for this study have been affected by this change in the law. Previous to December 15, 2003, firefighters who became ill as the result of a heart attack or stroke after going off duty needed to register a complaint of not feeling well while still on duty in order to be included in this study. For firefighter fatalities after December 15, 2003, firefighters will be included in this report if they became ill as the result of a heart attack or stroke within 24 hours of a training activity or emergency response. Firefighters who became ill after going off duty where the activities while on duty were limited to tasks that did not involve physical or mental stress will not be included.

A fatality may be caused directly by an accidental or intentional injury in either emergency or nonemergency circumstances, or it may be attributed to an occupationally related fatal illness. A common example of a fatal illness incurred on duty is a heart attack. Fatalities attributed to occupational illnesses also include a communicable disease contracted while on duty that proved fatal when the disease could be attributed to a documented occupational exposure.

Firefighter fatalities are included in this report even when death is considerably delayed after the original incident. When the incident and the death occur in different years, the analysis counts the fatality as having occurred in the year in which the incident took place.

There is no established mechanism for identifying fatalities that result from illnesses such as cancer that develop over long periods of time and which may be related to occupational exposure to hazardous materials or toxic products of combustion. It has proved to be very difficult to provide a complete evaluation of an occupational illness as a causal factor in firefighter deaths due to the following limitations: the exposure of firefighters to toxic hazards is not sufficiently tracked; the often delayed long-term effects of such toxic hazard exposures; and the exposures firefighters may receive while off duty.

Sources of Initial Notification

As an integral part of its ongoing program to collect and analyze fire data, the USFA solicits information on firefighter fatalities directly from the fire service and from a wide range of other sources. These sources include the PSOB Program administered by the DOJ, the National Institute for Occupational Safety and Health (NIOSH), the Occupational Safety and Health Administration (OSHA), the Department of Defense (DOD), the National Interagency Fire Center (NIFC), and other Federal agencies.

The USFA receives notification of some deaths directly from fire departments, as well as from such fire service organizations as the International Association of Fire Chiefs (IAFC), the International Association of Fire Fighters (IAFF), the NFPA, the National Volunteer Fire

Council (NVFC), State fire marshals, State fire training organizations, other State and local organizations, fire service Internet sites, news services, and fire service publications.

Procedure for Including a Fatality in the Study

In most cases, after notification of a fatal incident, initial telephone contact is made with local authorities by the USFA to verify the incident, its location, jurisdiction, and the fire department or agency involved. Further information about the deceased firefighter and the incident may be obtained from the chief of the fire department or designee over the phone or by other forms of data collection. After basic information is collected, a notice of the firefighter fatality is posted at the National Fallen Firefighters Memorial site in Emmitsburg, MD, the USFA website, and a notice of the fatality is transmitted by electronic mail to a large list of fire service organizations and fire service members.

Information that is requested routinely from fire departments that have experienced a fatality includes NFIRS-1 (incident) and NFIRS-3 (fire service casualty) reports; the fire department's own incident and internal investigation reports; copies of death certificates and autopsy results; special investigative reports; law enforcement reports; photographs and diagrams; and newspaper or media accounts of the incident. Information on the incident may also be gathered from NFPA or NIOSH reports.

After obtaining this information, a determination is made as to whether the death qualifies as an on-duty firefighter fatality according to the previously described criteria. With the exception of firefighter deaths after December 15, 2003, the same criteria were used for this study as in previous annual studies. Additional information may be requested by USFA, either through follow-up with the fire department directly, from State vital records offices, or other agencies. The final determination as to whether a fatality qualifies as an onduty death for inclusion in this statistical analysis is made by the USFA. The final determination as to whether a fatality qualifies as a line-of-duty death (LODD) for inclusion in the annual National Fallen Firefighters Memorial Service is made by NFFF.

2009 FINDINGS

Ninety (90) firefighters died while on duty in 2009, a substantial reduction in the number of firefighter fatalities over many previous years. This total includes 13 firefighters who died under circumstances as a result of inclusion criteria changes resulting from the Hometown Heroes Act of 2003. When not including these fatalities in a trend analysis, the 2009 total 77 firefighter fatalities equals the lowest number of firefighter losses on record (77 onduty firefighter deaths occurring in 1992) over the past 33 years.

An analysis of multiyear firefighter fatality trends needs to acknowledge the changes from the Hometown Heroes Survivors Benefit Act of 2003; therefore, some graphs and charts either will or will not indicate the Hometown Heroes portion of the total. However, this does not diminish the sacrifices made by any firefighter who dies while on duty or the sacrifices made by his/her family and peers.

Moreover, when conducting multiyear comparisons of firefighter fatalities in this report, the losses that were the result of the attacks on the World Trade Center (WTC) in New York City on September 11, 2001, are sometimes also set apart for illustrative purposes. This action is by no means a minimization of the supreme sacrifice made by these firefighters.

Figure 1. Onduty Firefighter Fatalities (1977-2009)

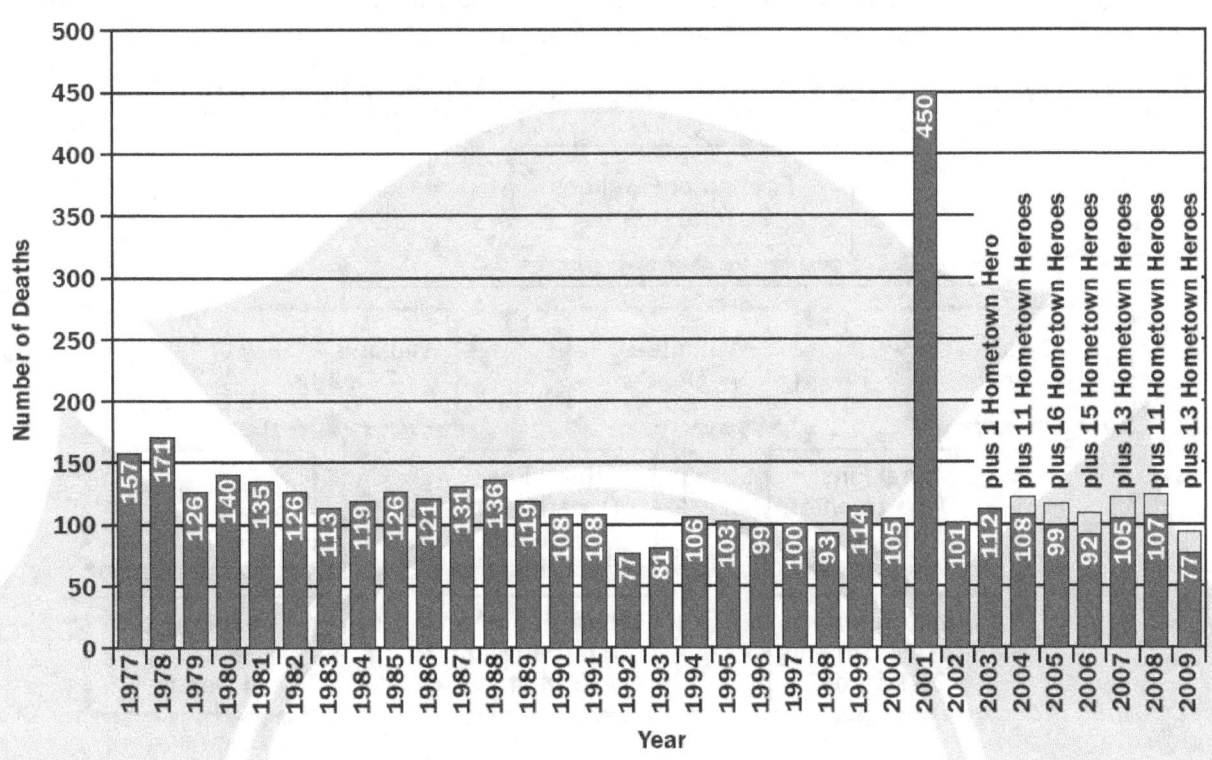

Figure 2. Firefighter Fatalities per 100,000 Fires

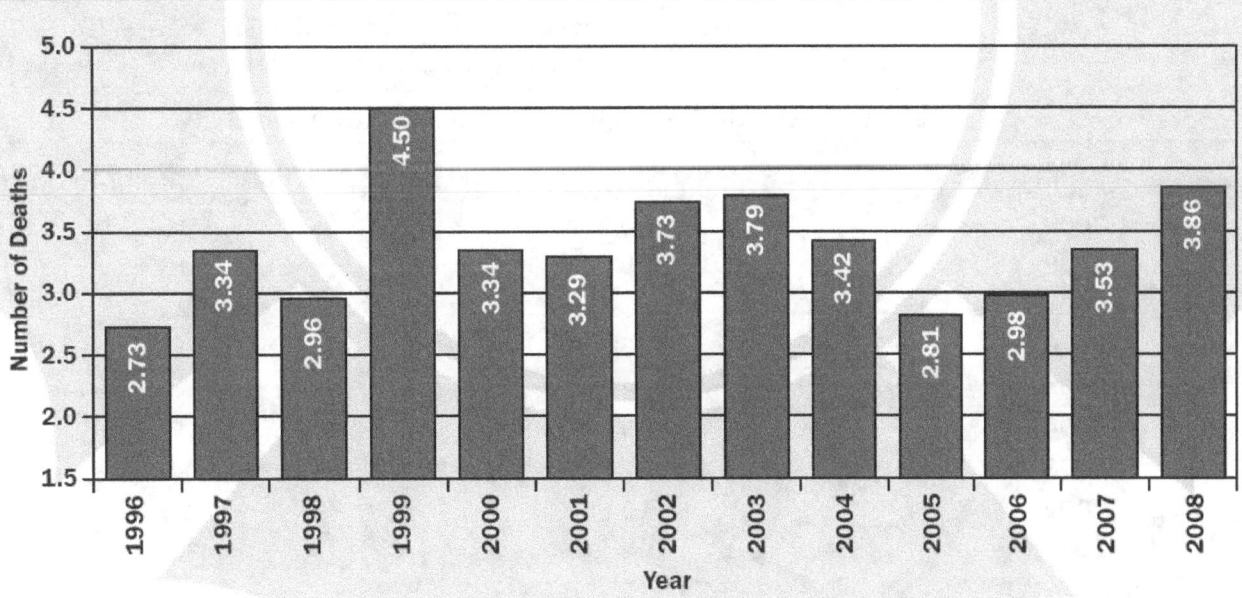

Career, Volunteer, and Wildland Agency Deaths

In 2009, firefighter fatalities included 47 volunteer firefighters, 36 career firefighters, and 7 part-time or full-time members of wildland or wildland contract fire agencies (Figure 3).

Figure 3. Career, Volunteer, and Wildland Agency Deaths (2009)

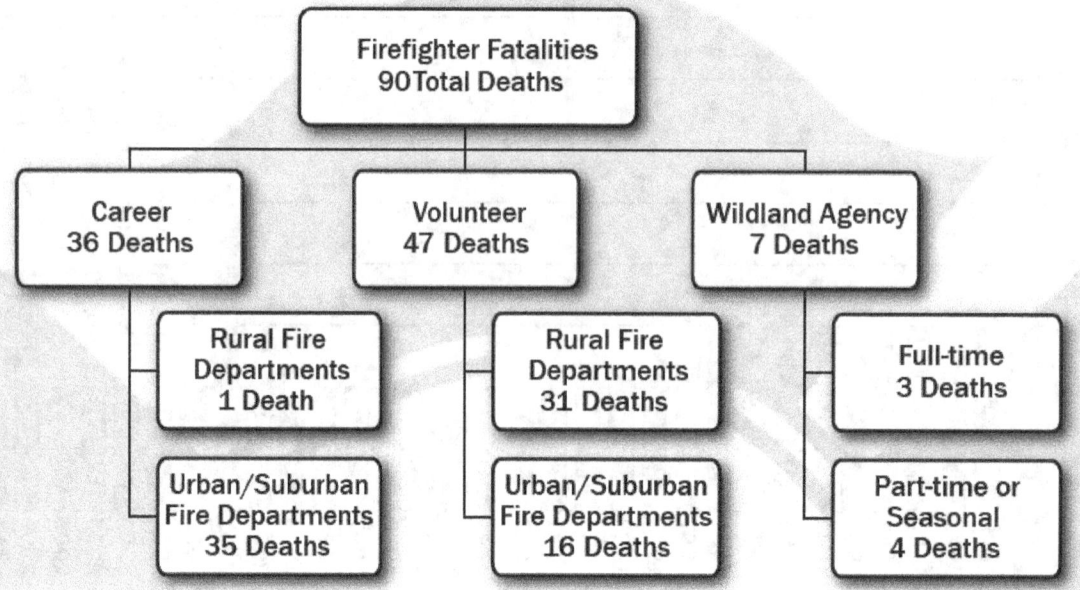

Gender

One of the firefighters who died in 2009 was female and 89 were male.

Multiple Firefighter Fatality Incidents

The 90 deaths in 2009 resulted from a total of 82 fatal incidents. There were 6 firefighter fatality incidents where 2 or more firefighters were killed in 2009, claiming a total of 13 firefighters.

Table 1. Multiple Firefighter Fatality Incidents

Year	Number of Incidents	Total Number of Deaths
2009	6	13
2008	5	18
2007	7	21
2006	6	17
2005	4	10
2004	3	6
2003	7	20
2002	9	25
2001	8	362
2001 w/o WTC	7	18
2000	5	10
1999	6	22
1998	10	22

Wildland Firefighting Deaths

In 2009, 16 firefighters were killed during activities involving brush, grass, or wildland firefighting. This total includes part-time and seasonal wildland firefighters, full-time wildland firefighters, and municipal or volunteer firefighters whose deaths are related to a wildland fire (Figure 4).

Figure 4. Firefighter Fatalities Related to Wildland Firefighting (1997-2009)

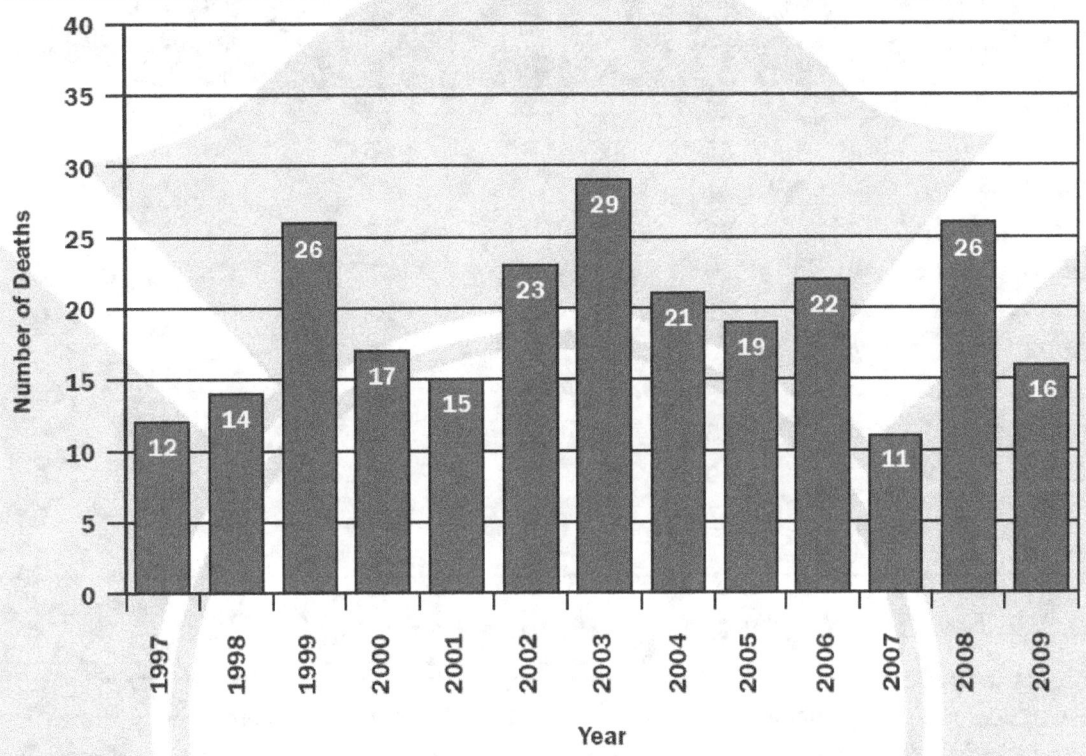

Table 2. Wildland Firefighting Aircraft Deaths

Year	Total Number of Deaths	Number of Fatal Incidents
2009	5	2
2008	16	4
2007	1	1
2006	8	3
2005	6	2
2004	3	3
2003	7	4
2002	6	3
2001	6	3
2000	6	5
1999	0	0
1998	3	2

In 2009, there were two multiple firefighter fatality incidents related to wildland firefighting that killed five firefighters.

Table 3. Firefighter Deaths Associated with Wildland Firefighting

Year	Total Number of Deaths	Number of Fatal Incidents	Number of Firefighters Killed in Multiple-Death Incidents
2009	16	13	5
2008	26	15	14
2007	11	11	0
2006	22	13	13
2005	19	15	6
2004	21	21	0
2003	30	22	10
2002	23	14	13
2001	15	9	9
2000	19	16	6
1999	27	26	2
1998	14	13	2

TYPE OF DUTY

Activities related to emergency incidents resulted in the deaths of 57 firefighters in 2009 (Figure 5). This includes all firefighters who died responding to an emergency or at an emergency scene, returning from an emergency incident, and during other emergency-related activities. Nonemergency activities accounted for 33 fatalities. Nonemergency duties include training, administrative activities, performing other functions that are not related to an emergency incident, and postincident fatalities where the firefighter does not experience the illness or injury during the emergency.

Figure 5. Firefighter Deaths by Type of Duty (2009)

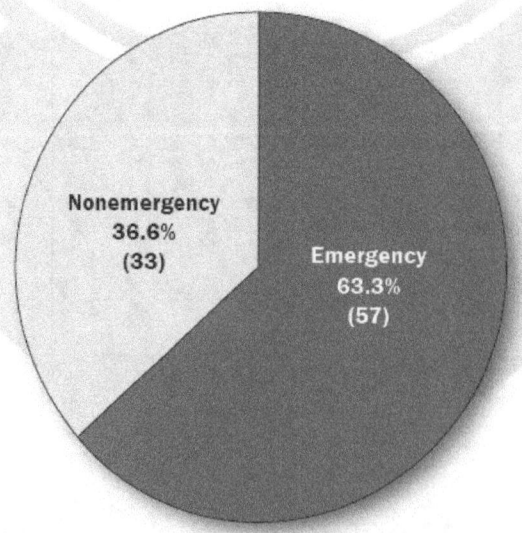

A multiyear historical perspective relating to the percentage of firefighter deaths that occurred during emergency duty is presented in Table 4. The percentage of all deaths for 2004 and the years after are lower for each year due to the inclusion of firefighters covered by the changes resulting from the Hometown Heroes Act of December 2003. As such, a second column has been added to Table 4 for purposes of a longer-term trend analysis.

Table 4. Emergency Duty Firefighter Deaths

Year	Percentage of All Deaths	Percentage of All Deaths Without Hometown Heroes
2009	63.3	82.2
2008	63.5	70
2007	64.4	72.4
2006	57.5	66.3
2005	52.1	60.6
2004	68.9	75.9
2003	69	69.6
2002	73	73
2001	65	65
2001 with WTC	92	92
2000	71	71
1999	87	87
1998	77	77

In 2009, 49 firefighters died while responding to or working on the scene of an emergency. This number includes deaths resulting from injuries sustained on the incident scene or en route to the incident scene and firefighters who became ill on an incident scene and later died. It does not include firefighters who became ill or died after or while returning from an incident, e.g., a vehicle collision. Figure 6 shows the number of firefighters killed while responding to or working on the scene of an emergency in 2009.

Thirty-nine firefighters were killed during firefighting duties; 3 firefighters were killed on emergency medical services (EMS) calls; 5 on motor vehicle accidents; 1 firefighter was killed in association with a weather incident; and 1 was killed during other emergency circumstances.

Figure 6. Type of Emergency Duty

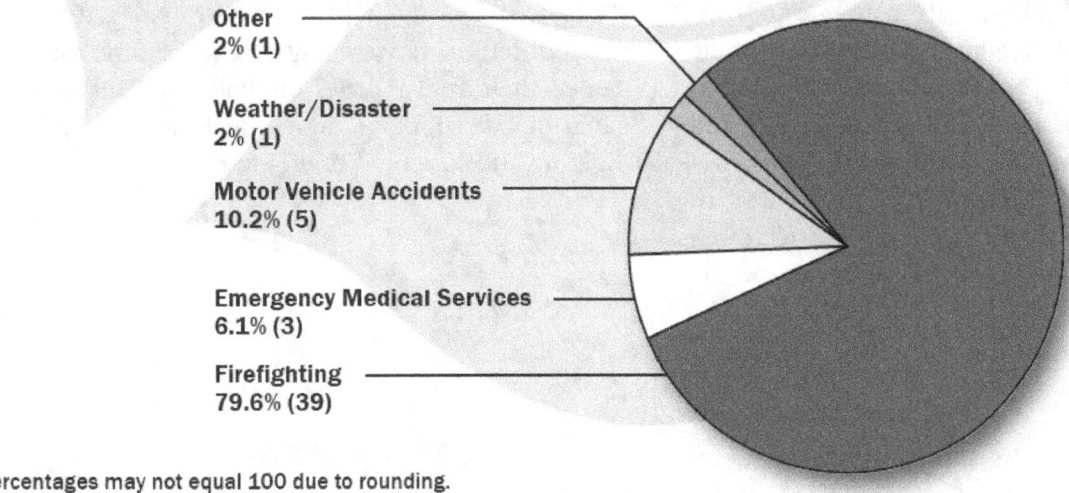

Other
2% (1)

Weather/Disaster
2% (1)

Motor Vehicle Accidents
10.2% (5)

Emergency Medical Services
6.1% (3)

Firefighting
79.6% (39)

Percentages may not equal 100 due to rounding.

The number of deaths by type of duty being performed in 2009 is shown in Table 5 and presented graphically in Figure 7.

Table 5. Firefighter Deaths by Type of Duty (2009)

Type of Duty	Number of Deaths
Fireground Operations	30
Responding/Returning	15
Other Onduty Deaths	12
Training	10
Nonfire Emergencies	9
After an Incident	14
Total	90

Figure 7. Firefighter Deaths by Type of Duty (2009)

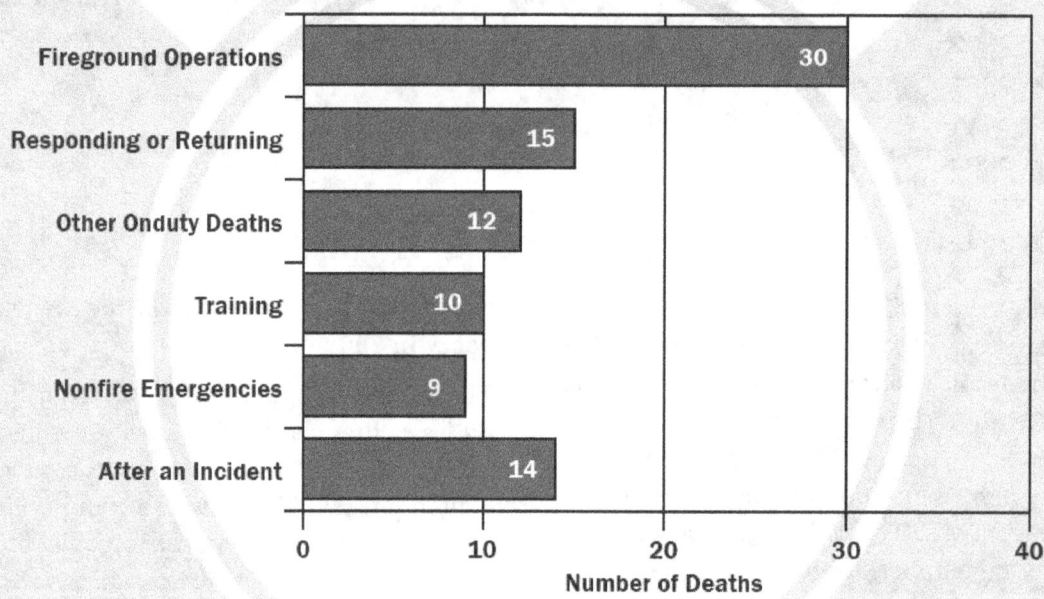

Fireground Operations

Of the 30 firefighters killed during fireground operations in 2009, 19 firefighters died while on the scene of a structure fire, 9 firefighters died while en route or at the scene of a wildland or outside fire, and 1 firefighter at the scene of a vehicle fire. One other firefighter fell ill while at the scene of an alarm in an apartment building and later died from a cerebrovascular accident (CVA) after being transported to the hospital.

Type of Fireground Activity

Table 6 shows the types of fireground activities in which firefighters were engaged at the time they sustained their fatal injuries or illnesses. This total includes all firefighting duties, such as wildland firefighting and structural firefighting.

Table 6. Type of Activity (2009)

Water Supply	2
Pump Ops	4
Search & Rescue	4
Advance Hoselines	12
Support	2
Other	1
Unknown	4
Scene Safety	1

Fixed Property Use for Structural Firefighting Deaths

There were 19 fatalities in 2009 where firefighters became ill or injured while on the scene of a structure fire. Table 7 shows the distribution of these deaths by fixed property use. As in most years, residential occupancies accounted for the highest number of these fireground fatalities in 2009.

Table 7. Structural Firefighting Deaths by Fixed Property Use in 2009

Residential	13
Commercial	6

Responding/Returning

Fifteen firefighters died while responding to or returning from 13 emergency incidents in 2009; 11 while responding and 4 while returning. This compares to 24 responding/returning firefighter fatalities in 2008. Two of the responding firefighter deaths did not involve motor vehicles. One of the incidents, an air tanker crash, took the lives of three firefighters.

The types of fire department apparatus involved in vehicle collisions killing six firefighters included one engine, one tanker (tender), one ladder, one ambulance, one brush truck, and one staff/support vehicle. Four of the vehicle collisions occurred while responding and two while returning.

The status of seatbelt use in fire apparatus collisions was reported in all but one instance where the firefighter/passenger of an ambulance was ejected from the vehicle (likely not to have been wearing a seatbelt) and killed when responding to an incident. In four of the remaining five fire apparatus collisions, no seatbelt was being worn by the firefighter killed, two drivers and two passengers respectively, and two of these were ejected completely from their apparatus (one driver and one passenger). Only one firefighter, who responded from his residence in a fire department pickup truck, was reported to have been wearing his seatbelt when another vehicle at an intersection pulled into his path causing his truck to leave the narrow road and strike a tree. The firefighter was not ejected from the vehicle, but passed away at the scene from multiple traumatic injuries.

Speed was not reported to have been a factor in any of the six responding/returning fire apparatus collisions, but four occurred at an intersection (two of these at night).

One firefighter died while returning from an incident as the sole occupant and operator of an engine. While driving, he experienced a heart attack causing the apparatus to leave the roadway and travel across a field coming to a stop against a camper.

Three firefighters were killed while responding to or returning from emergency incidents in their privately-owned vehicles (POVs). Two of these deaths were caused by heart attacks, one while responding, where the firefighter pulled to the side of the road, and the other while returning from an incident, where the firefighter's POV left the roadway and crashed. The third firefighter lost control of his POV, crossed the center line, left the roadway, and struck a tree. The 18-year old firefighter was killed as a result of massive head trauma sustained during the crash. He was not wearing a seatbelt at the time of the crash and was reported to have been fully ejected from his vehicle. Speed and wet road conditions were cited as factors in the crash, which occurred late at night on a curved section of roadway.

Table 8. Firefighter Deaths While Responding to or Returning from an Incident

Year	Number of Firefighter Deaths
2009	15
2008	24
2007	26
2006	15
2005	22
2004	23
2003	36
2002	13
2001	23
2000	19
1999	26
1998	14

Training

In 2009, 10 firefighters died while engaged in training activities. Three of the deaths were due to heart attacks, two from CVA/strokes, one from heat exhaustion, three from falls, and one from being struck by an object.

Table 9. Firefighter Fatalities While Engaged in Training

Year	Number of Firefighter Deaths
2009	10
2008	12
2007	11
2006	9
2005	14
2004	13
2003	12
2002	11
2001	14
2000	13
1999	3
1998	12

Nonfire Emergencies

In 2009, there were nine firefighter fatalities where the type of emergency duty was not related to a fire. Four were from motor vehicle accidents, four from EMS incidents, and one fatality was related to an inclement weather incident.

After the Incident

In 2009, 14 firefighters died after the conclusion of their onduty activity. Six deaths were due to heart attacks, five were due to CVA/strokes, and three were due to other causes (one aortic separation, one from asthma, and one unknown).

CAUSE OF FATAL INJURY

The term "cause of injury" refers to the action, lack of action, or circumstances that resulted directly in the fatal injury. The term "nature of injury" refers to the medical cause of the fatal injury or illness which is often referred to as the physiological cause of death. A fatal injury usually is the result of a chain of events, the first of which is recorded as the cause.

Figure 8 shows the distribution of deaths by cause of fatal injury or illness in 2009.

Figure 8. Fatalities by Cause of Fatal Injury (2009)

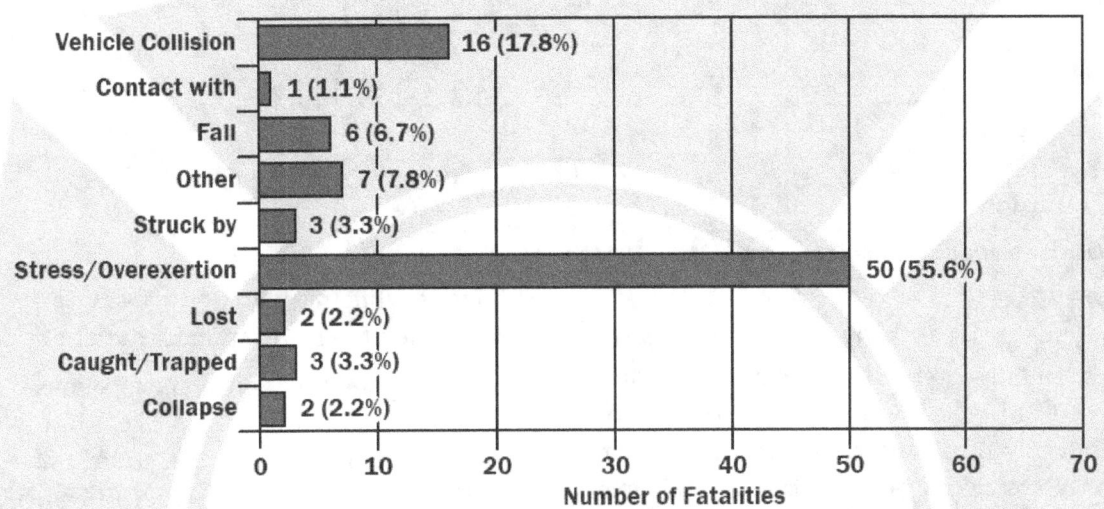

Stress or Overexertion

Firefighting is extremely strenuous physical work and is likely one of the most physically demanding activities that the human body performs.

Stress or overexertion is a general category that includes all firefighter deaths that are cardiac or cerebrovascular in nature such as heart attacks, strokes, and other events such as extreme climatic thermal exposure. Classification of a firefighter fatality in this cause of fatal injury category does not necessarily indicate that a firefighter was in poor physical condition.

Fifty firefighters died in 2009 as a result of stress/overexertion:

• Thirty-nine firefighters died due to a heart attack.

• Eight firefighters died due to CVAs.

• One firefighter died from heat exhaustion.

• One firefighter died from a pulmonary embolism.

• One firefighter died from damage to a heart valve, an acute event caused by the extreme physical exertion.

Table 10. Deaths Caused by Stress or Overexertion

Year	Number	Percent of Fatalities
2009	50	55.5
2008	52	44
2007	55	46.6
2006	54	50.9
2005	62	53.9
2004	66	56.4
2003	51	45.9
2002	38	38
2001	43	40.9*
2000	46	44.6
1999	56	49.5
1998	43	46.2

Does not include the firefighter deaths of September 11, 2001, in New York City.

Vehicle Crashes

Sixteen firefighters died in 2009 as the result of vehicle crashes, slightly over half the number of firefighter deaths from vehicle crashes in 2008. Five of these deaths involved aircraft, only two occurred in POVs (the POV average 2004-2008 is 7.6), and nine involved fire department apparatus.

- Of the two crashes that involved the firefighter's personal vehicle, one death occurred while the firefighter was returning from training on his motorcycle and struck a deer. The second firefighter died from injuries while responding to a vehicle fire when his vehicle left the roadway and struck a tree.

- Nine separate fire department apparatus crashes took the lives of nine firefighters. These involved two engines, one tanker, one ladder truck, one support vehicle, one ambulance, and three brush trucks. Of the nine apparatus accidents, the seatbelt status is known for five. Four of these five accidents involved firefighters who were not wearing seatbelts and two of these firefighters were fully ejected from their vehicles.

Figure 9. Firefighter Fatalities in Vehicle Collisions

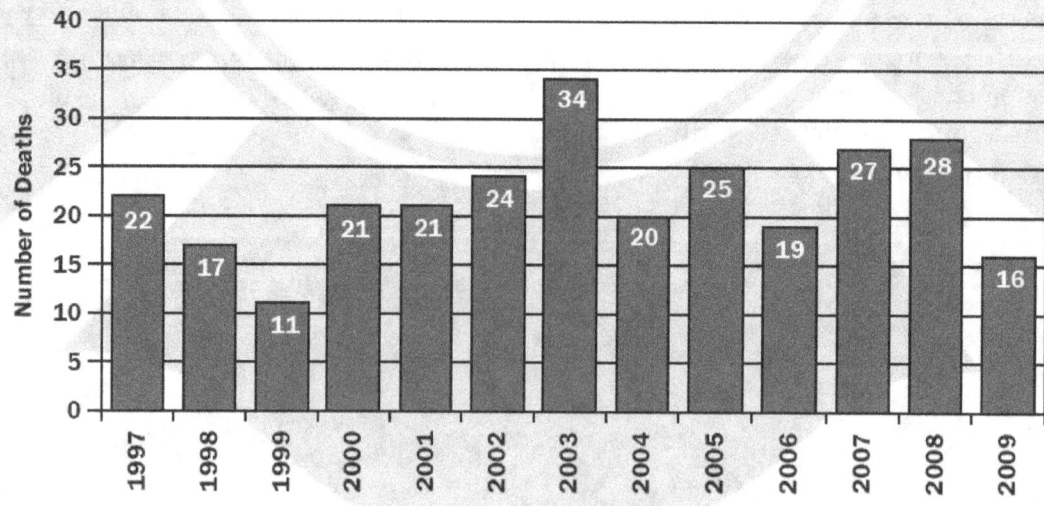

Lost or Disoriented

Two firefighters died in 2009 when they became lost or disoriented inside of a manufactured home next to a camper where the fire had originated. The firefighters advanced an attack line into the home as other firefighters attacked the fire in the camper. Five to 10 minutes after their entry, the pump operator sounded an evacuation signal, concerned that he was running out of water. When the two firefighters did not emerge from the home, firefighters called out for them, attempted to contact them on the radio, and tugged on the attack line to no avail. The firefighters were eventually discovered in the front room of the home unconscious. Both firefighters were pronounced dead at the scene.

Caught or Trapped

Three firefighters were killed in 2009 in two separate incidents when they were caught or trapped. This classification covers firefighters trapped in wildland and structural fires who were unable to escape due to rapid fire progression and the byproducts of smoke, heat, toxic gases, and flame. This classification also includes firefighters who drowned, and those who were trapped and crushed.

- The cause of death for one firefighter was listed as asphyxiation due to probable carbon monoxide toxicity after he had re-entered a large grain silo to assist a fellow firefighter attempt an exit from the structure. Both firefighters subsequently lost consciousness. Firefighters on the exterior cut a hole in the metal wall of the bin and extricated the firefighters, saving one.

- Two firefighters were caught and trapped after they advanced an attack line to the interior of the residence and fire conditions changed rapidly.

Collapse

Two firefighters died in 2009 while they were searching a burning commercial structure and the main floor collapsed trapping the firefighters.

Struck by Object

Being struck by an object was the cause of three fatal firefighter injuries in 2009.

- One firefighter was shot and killed by a patient who became violent while being treated by emergency services personnel.

- One firefighter was struck and killed by a vehicle on a roadway where he was working to clear trees downed by inclement weather.

- One firefighter was struck and killed by a tree while working a hazard tree abatement project.

Fall

Six firefighters died in 2009 as the result of falls.

- Two firefighters fell from a new 95-foot midmount aerial ladder tower.

- One firefighter fell from a fire department ladder truck following a parade.

- One firefighter suffered a serious head injury sustained from a fall in the firehouse.

- One firefighter fell while performing routine rappel proficiency skill training.

- One of three firefighters did not survive a fall when escaping rapidly progressing fire conditions forcing them to jump from the third story of a residence.

Contact With

One firefighter was fatally electrocuted while at the scene of a motor vehicle accident where at least one power line was brought to the ground by the crash and others were drooping closer to the ground. When attempting to step over a power line on the ground, the firefighter stumbled, attempted to regain his balance, and came into contact with one or more power lines.

Other

Seven firefighters died in 2009 of a cause that is not categorized above.

- One firefighter slipped on the driveway at the scene of a medical emergency incident and sustained an injury to his leg that required surgery. As the firefighter recovered from his surgery, he suffered a blood clot in his lungs (pulmonary embolism). The autopsy indicated that the clot was related to his leg injury.

- While at the scene of a structure fire, one firefighter complained of a headache and told others that he could not see out of one eye. He was transported to the hospital. Tests at the hospital revealed a colloid cyst in his brain that several days later resulted in the firefighter's death.

- One firefighter positioned the department's tower ladder truck at the rear of the fire station. He set the stabilizers and entered the ladder's platform area. The firefighter then maneuvered the platform into the open rear apparatus bay door of the fire station. His head became trapped between the door frame and the platform railing causing fatal injuries.

- One firefighter returned to his fire station after responding to a residential carbon monoxide alarm. At the station, the firefighter began complaining of chest pains and was subsequently transported to the hospital. After undergoing surgery for an aortic separation, he went into cardiac arrest and died the following morning.

- One fire police officer worked traffic control at the scene of a mutual-aid vehicle crash with entrapment. He cleared the scene and went home. Within 2 hours of leaving the scene of the incident, the fire police officer died as the result of illness related to asthma.

- One firefighter passed away at home after responding to a series of EMS and fire calls. The cause and nature of his fatal injury are unknown.

- One firefighter died as a result of injuries sustained at a manufacturing company when firefighters applied water to a burning dumpster and a large explosion occurred.

NATURE OF FATAL INJURY

Figure 10 shows the distribution of the 90 firefighter deaths that occurred in 2009 by the medical nature of the fatal injury or illness. For heart attacks, Figure 11 shows the type of duty involved.

Figure 10. Fatalities by Nature of Fatal Injury (2009)

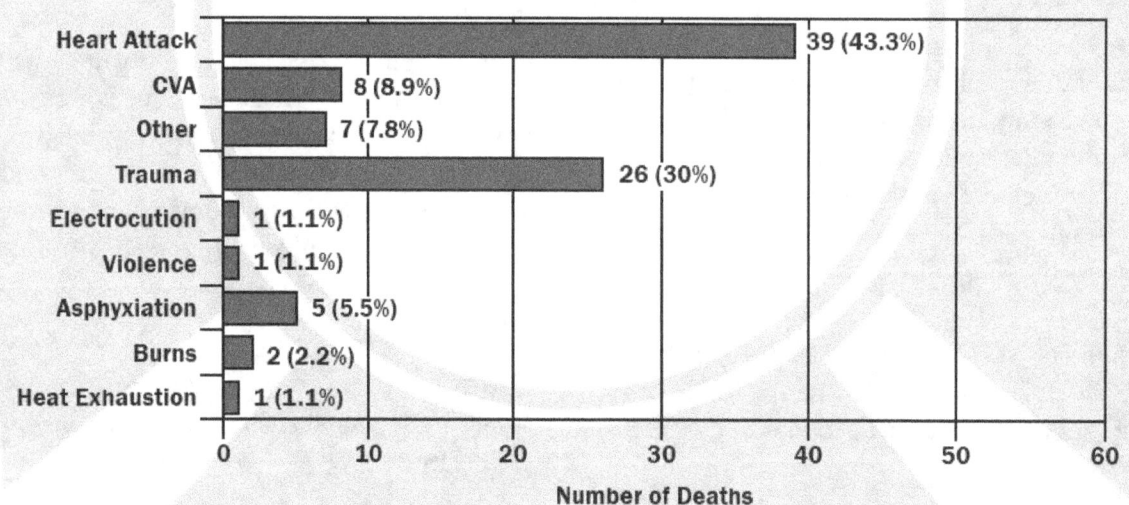

Figure 11. Heart Attacks by Type of Duty (2009)

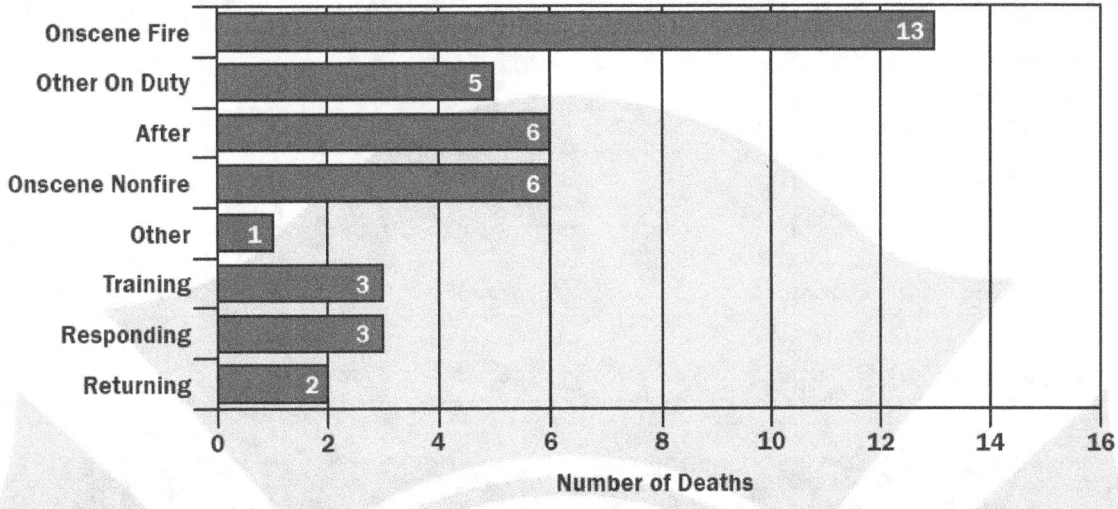

FIREFIGHTER AGES

Figure 12 shows the percentage distribution of firefighter deaths by age and nature of the fatal injury. Table 11 provides a count of firefighter fatalities by age and the nature of the fatal injury.

Younger firefighters were more likely to have died as a result of traumatic injuries such as injuries from an

apparatus accident or becoming caught or trapped during firefighting operations. Stress-related deaths are rare below the 31 to 35 years of age category and, when they occur, often include underlying medical conditions.

Figure 12. Fatalities by Age and Nature (2009)

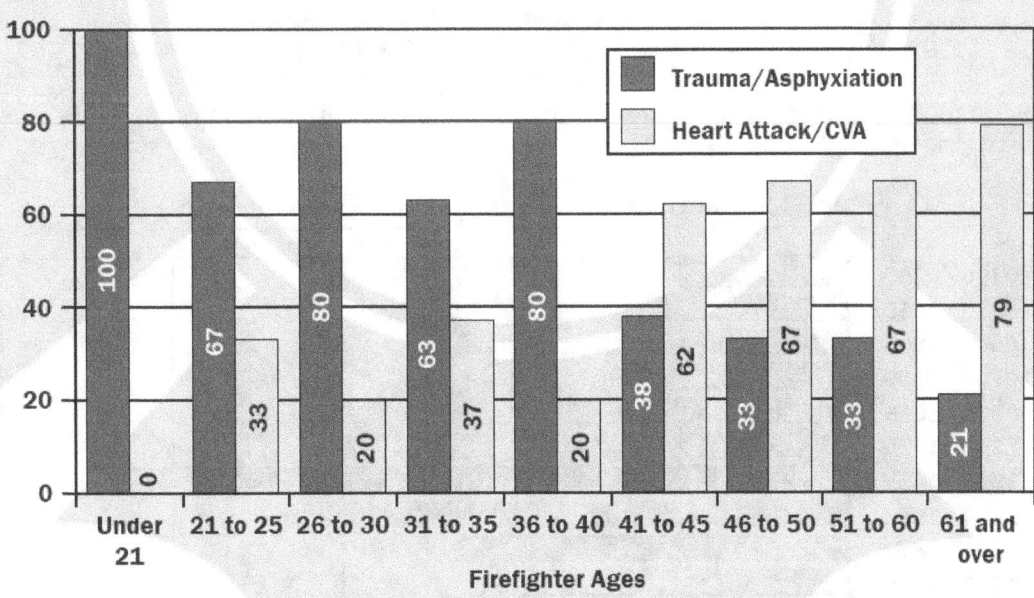

Table 11. Firefighter Ages and Nature of Fatal Injury (2009)

Age Range	Heart Attack/CVA/Other	Trauma/Asphyxiation Total
under 21	0	2
21 to 25	1	2
26 to 30	1	4
31 to 35	3	5
36 to 40	1	4
41 to 45	10	6
46 to 50	6	3
51 to 60	14	7
61 and over	11	3

The youngest firefighter to die in 2009 was age 18. He was responding to a car fire in his personal vehicle when he lost control of the vehicle, crossed the center line, left the roadway, and struck a tree. The firefighter was killed as a result of the crash due to massive head trauma. He was not wearing a seatbelt at the time of the crash. Speed and wet road conditions were cited as factors in the crash.

The oldest firefighter killed on duty in 2009 was 77. He died at the fire station of a heart attack shortly after preparing vehicles and assisting other firefighters responding to a reported wildland fire.

DEATHS BY TIME OF INJURY

The distribution of all 2009 firefighter deaths according to the time of day when the fatal injury occurred is illustrated in Figure 13. The time of fatal injury for 11 firefighters was either unknown or not reported.

Figure 13. Fatalities by Time of Fatal Injury (2009)

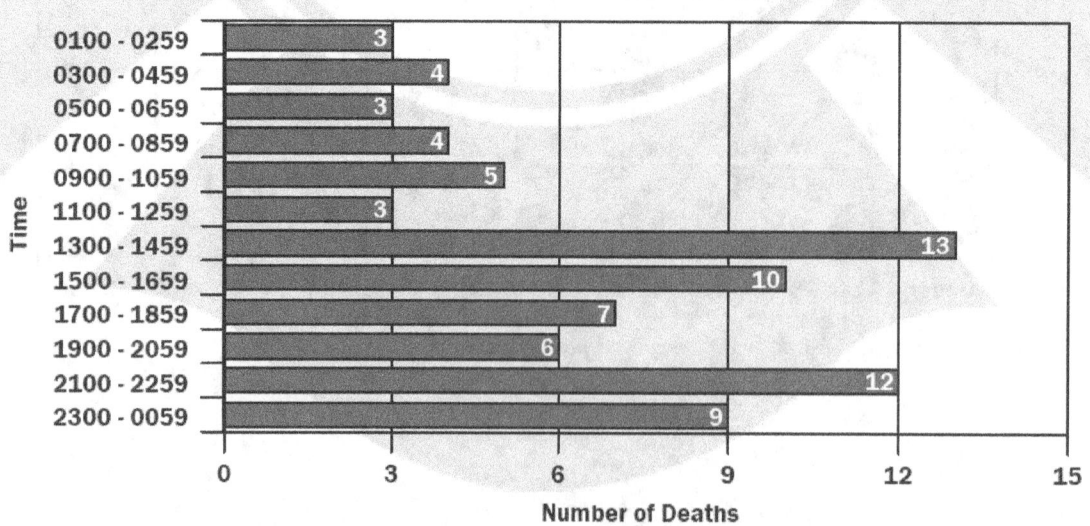

FIREFIGHTER FATALITY INCIDENTS BY MONTH OF YEAR

Figure 14 illustrates the 2009 firefighter fatalities by month of the year.

Figure 14. Deaths by Month of Year (2009)

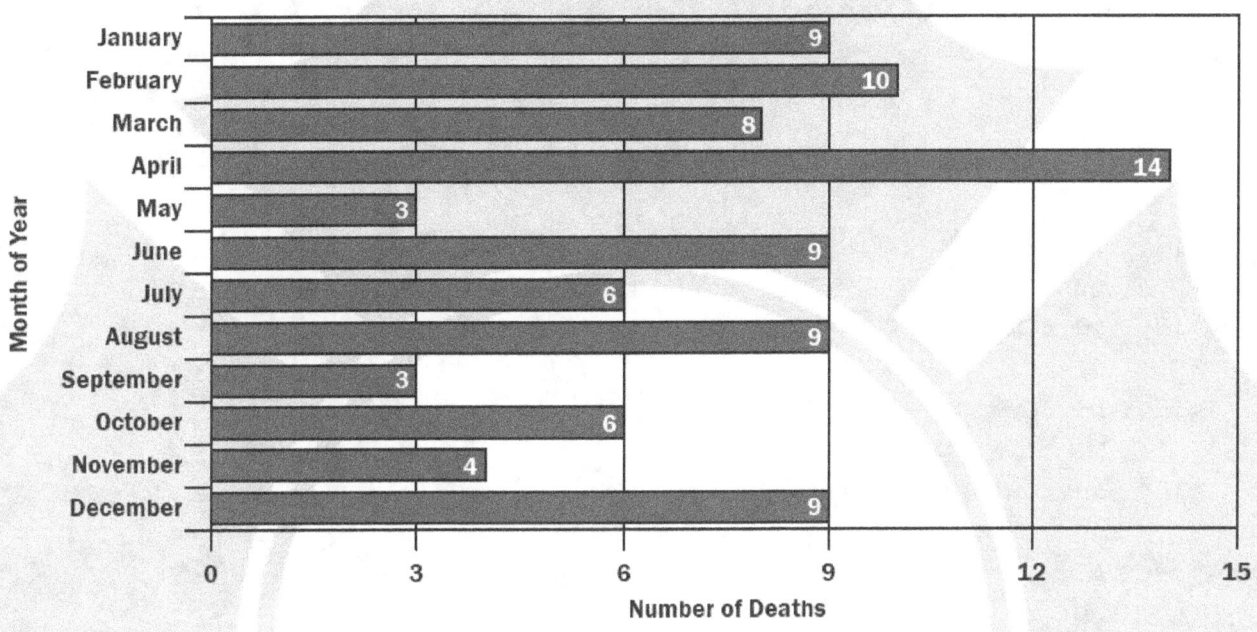

STATE AND REGION

The distribution of firefighter deaths in 2009 by State is shown in Table 12. Firefighters based in 32 States died in 2009.

The highest number of firefighter deaths, based on the location of the fire service organization in 2009, oc-curred in Pennsylvania with eight deaths. New York and North Carolina had the next highest totals of firefighter fatalities in 2009 respectively, seven and six firefighter deaths each.

Table 12. Firefighter Fatalities by State by Location of Fire Service* (2009)

State	Fatalities	Percentage
AL	2	2.2
AZ	1	1.1
CA	3	3.3
CO	1	1.1
CT	2	2.2
FL	2	2.2
GA	3	3.3
IL	2	2.2
IN	1	1.1
KS	3	3.3
KY	1	1.1
LA	5	5.6
MA	2	2.2
MD	1	1.1
ME	1	1.1
MO	3	3.3
MS	3	3.3
MT	4	4.4
NC	6	6.7
NJ	2	2.2
NY	7	7.8
OH	3	3.3
OK	4	4.4
PA	8	8.9
RI	1	1.1
SC	1	1.1
TX	5	5.6
UT	2	2.2
VA	3	3.3
VT	2	2.2
WI	4	4.4
WV	2	2.2

* This list attributes the deaths according to the State in which the fire department or unit is based, as opposed to the State in which the death occurred. They are listed by those States for statistical purposes and for the National Fallen Firefighters Memorial at the NETC. Due to rounding, percentage totals will not add to 100.

Figure 15. Firefighter Fatalities by Region (2009)

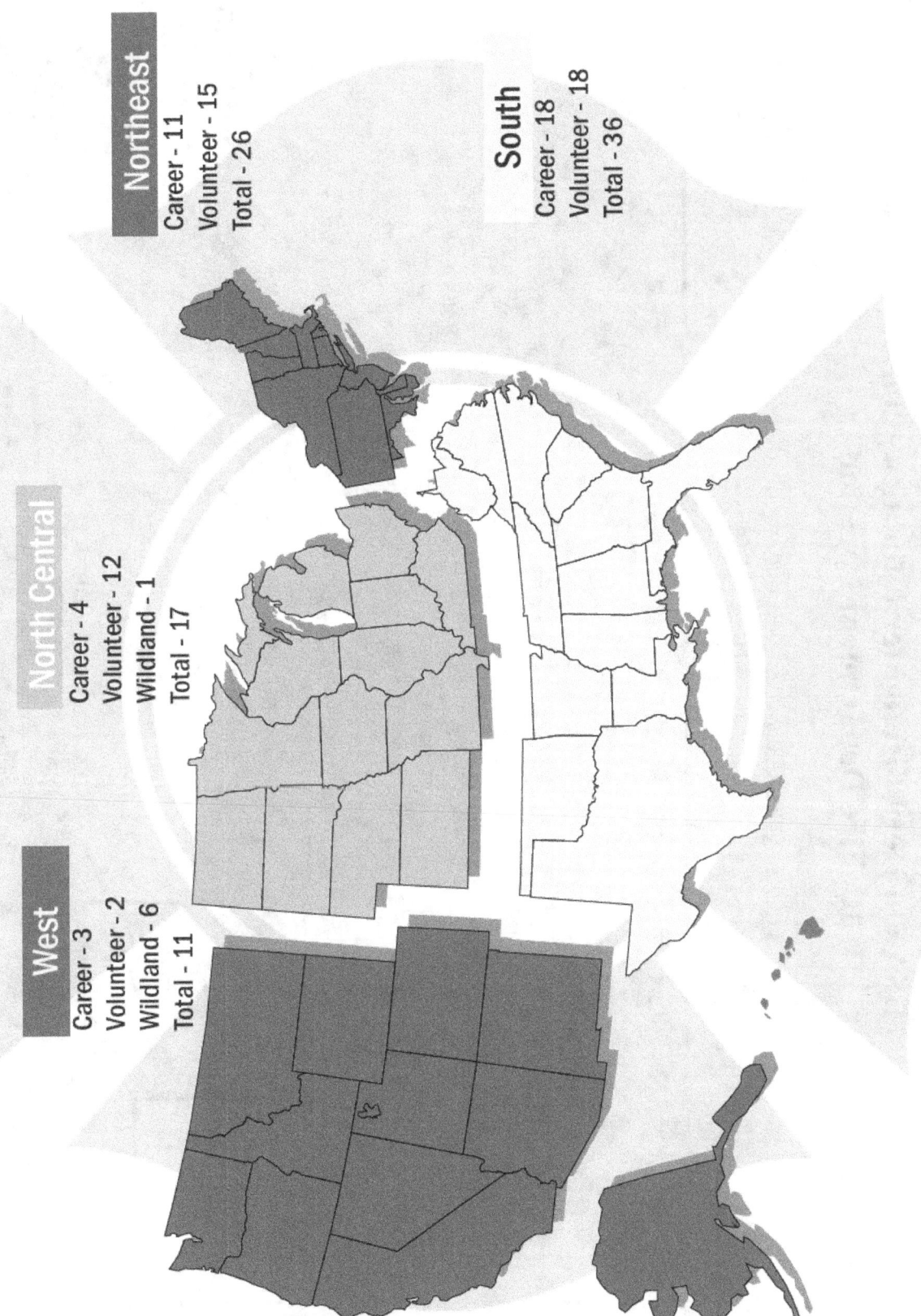

Northeast
Career - 11
Volunteer - 15
Total - 26

South
Career - 18
Volunteer - 18
Total - 36

North Central
Career - 4
Volunteer - 12
Wildland - 1
Total - 17

West
Career - 3
Volunteer - 2
Wildland - 6
Total - 11

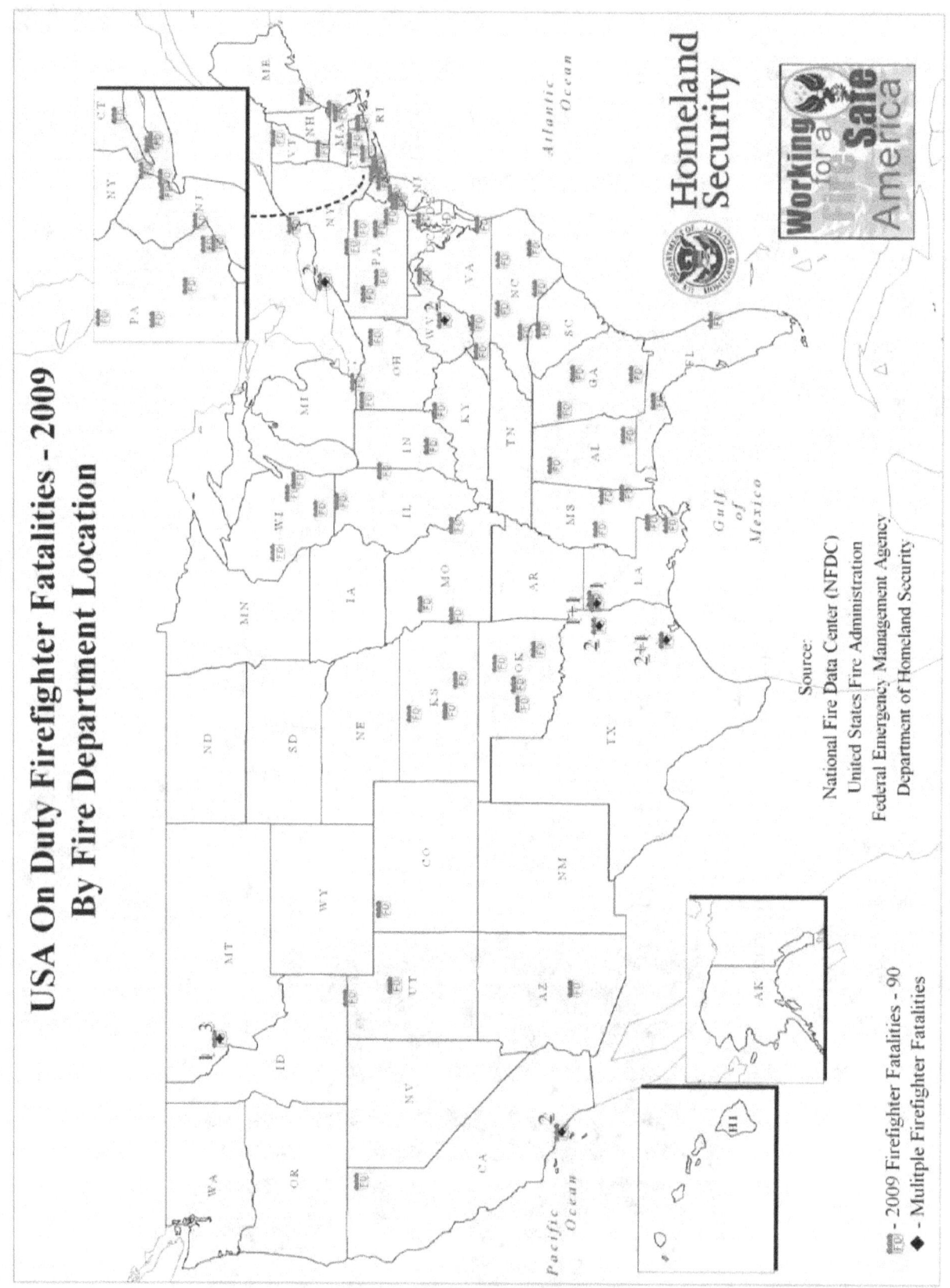

Figure 17. Onduty Firefighter Fatalities 2009 by Incident Location

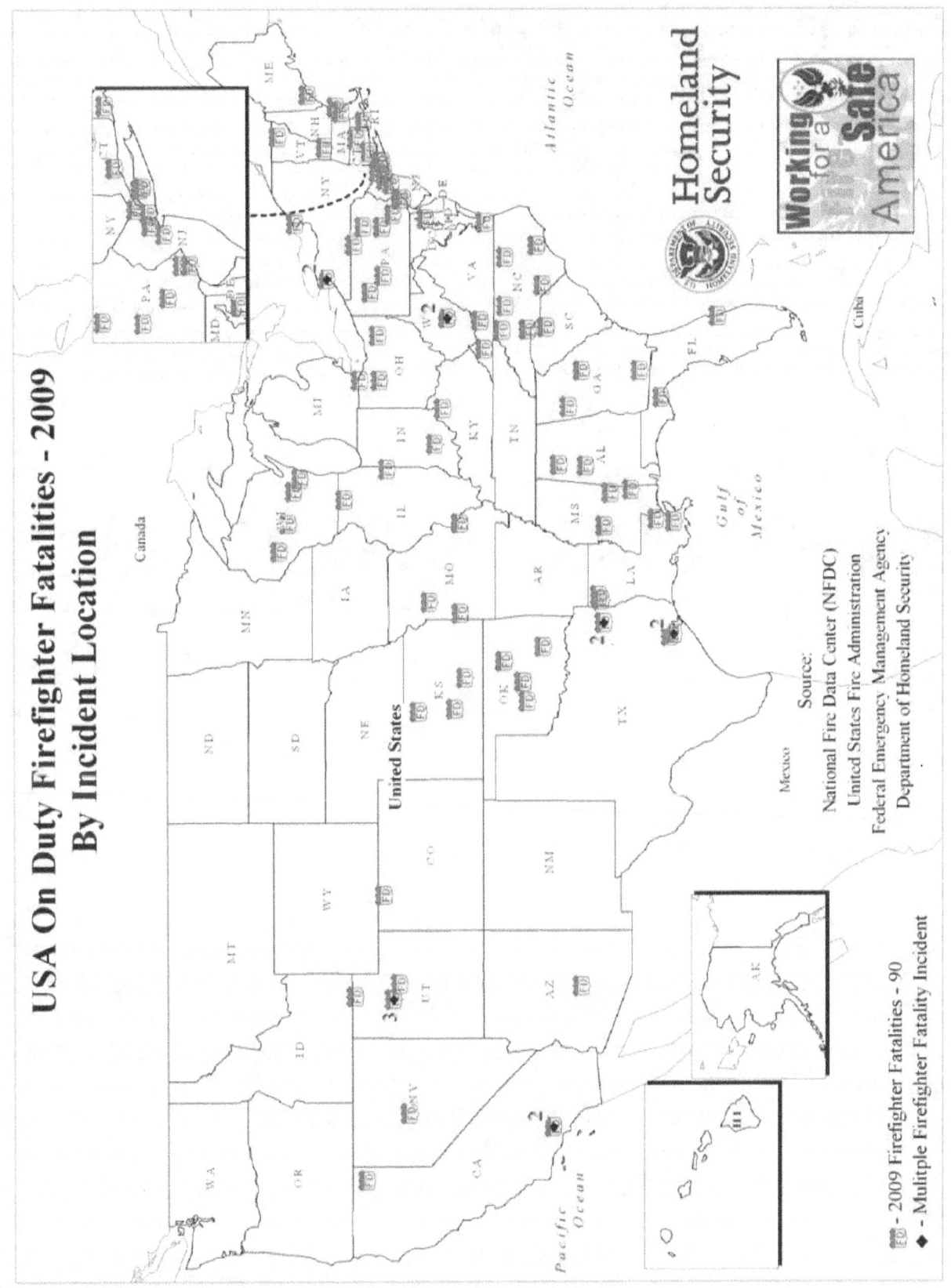

USA On Duty Firefighter Fatalities – 2009
By Incident Location

- 2009 Firefighter Fatalities

Source:
National Fire Data Center (NFDC)
United States Fire Administration
Federal Emergency Management Agency
Department of Homeland Security

Homeland Security

- 2009 Firefighter Fatalities - 90
- Mulitple Firefighter Fatality Incident

ANALYSIS OF URBAN/RURAL/SUBURBAN PATTERNS IN FIREFIGHTER FATALITIES

The U.S. Census Bureau defines "urban" as a place having a population of at least 2,500 or lying within a designated urban area. "Rural" is defined as any community that is not urban. "Suburban" is not a census term but may be taken to refer to any place, urban or rural, that lies within a metropolitan area defined by the Census Bureau, but not within one of the central cities of that metropolitan area.

Fire department areas of responsibility do not always conform to the boundaries used by the Census Bureau. For example, fire departments organized by counties or special fire protection districts may have both urban and rural coverage areas. In such cases, where it may not be possible to characterize the entire coverage area of the fire department as rural or urban, firefighter deaths were listed as urban or rural based on the particular community or location in which the fatality occurred.

The following patterns were found for 2009 firefighter fatalities. These statistics are based on answers from the fire departments and, when no data from the departments were available, the data were based upon population and area served as reported by the fire departments.

Table 13. Firefighter Deaths by Coverage Area Type (2009)

	Urban/Suburban	Rural	Federal or State Parks/Wildland	Total
Firefighter Deaths	51	32	7	90

In memory of all firefighters

who answered their last call in 2009

To their families and friends

To their service and sacrifice

January 2, 2009–0213 hrs
Gary Vernon Stephens, Firefighter
Age 57, Career • Elizabeth Fire Department, New Jersey

Firefighter Stephens and the members of his engine company were dispatched with other firefighters to a report of a structure fire. Firefighter Stephens was the acting captain of his company. The first-due company observed fire from a distance and requested a second alarm.

When Firefighter Stephens' company arrived on the scene, they were directed to establish a water supply for a ladder that was already on the scene. In order to access the scene, the engine was to back down the street where the incident was located, drop a supply line, and lay out to a hydrant. The street was crowded with parked vehicles. Firefighter Stephens dismounted the engine and walked behind it to guide the apparatus driver.

As the engine backed up, Firefighter Stephens walked behind it from the officer's side to the driver's side. As he walked, Firefighter Stephens was facing away from the moving apparatus. Firefighter Stephens was struck and run over by the apparatus.

Firefighters at the scene witnessed the incident and quickly removed Firefighter Stephens from under the apparatus. He was treated on the scene and transported to the hospital. He was pronounced dead at 0240 hours. The cause of death was listed as multiple blunt trauma injuries.

A homeless 19-year-old man was charged with setting the fire in the vacant residence as a warming fire.

For additional information regarding this incident, please refer to NIOSH Fire Fighter Fatality Investigation and Prevention Program Report F2009-10 (http://www.cdc.gov/niosh/fire/reports/face200910.html).

January 3, 2009–1533 hrs
John Covington "JC" Myers, Firefighter/Mechanic
Age 61, Volunteer • Union Chapel Volunteer Fire Department, Oklahoma

Firefighter/Mechanic Myers was operating a fire department brush truck at the scene of a wildland fire. The brush truck was being driven on a gravel road in heavy smoke conditions.

As the brush truck proceeded on the gravel road, it was involved in a head-on crash with a private pickup truck operated by another firefighter that had crossed the center of the road. Firefighter/Mechanic Myers was trapped in the brush truck for more than 90 minutes as a result of the crash.

Firefighter/Mechanic Myers died at the scene as a result of his injuries. The firefighter and another firefighter passenger in the other vehicle were treated for non-life threatening injuries. No seatbelts were in use in either vehicle.

January 9, 2009–1430 hrs
Kevin M. Kelley, Lieutenant
Age 52, Career • Boston Fire Department, Massachusetts

Lieutenant Kelley and his ladder company had responded to an emergency medical incident. After the incident concluded, Lieutenant Kelley and the members of his crew were returning to quarters.

As their apparatus descended a steep hill, the unit's brakes failed. When he realized that the apparatus could not be stopped, Lieutenant Kelley sounded the apparatus warning devices to warn pedestrians and other drivers and ordered his crew to brace for impact.

The ladder truck reached the bottom of the hill, crossed several lanes of traffic, crashed through several parked vehicles and an 8-foot brick fence, and came to rest embedded in a highrise building. Lieutenant Kelley was killed in the crash. The driver and two firefighter passengers in the ladder truck were injured. The driver had to be extricated from the damaged cab of the apparatus.

Investigations after the crash revealed serious deficiencies in the management and maintenance of the Boston Fire Department fleet.

NIOSH developed a safety advisory related to the maintenance of automatic slack adjusters on brake systems (2010-102). This advisory can be found at http://www.cdc.gov/niosh/fire/SafetyAdvisory10202009.html

For additional information regarding this incident, please refer to NIOSH Fire Fighter Fatality Investigation and Prevention Program Report F2009-05 (http://www.cdc.gov/niosh/fire/reports/face200905.html).

A Board of Inquiry Report prepared by the Boston Fire Department can be downloaded at http://www.cityofboston.gov/fire/pdfs/BoardofInquiryReport_Kelley.pdf

January 24, 2009–0230 hrs
Richard Lynn Rhea, Captain
Age 60, Volunteer • Crawfordville Volunteer Fire Department, Florida

A pickup truck was involved in a crash that resulted in injuries to the occupants of the truck and a broken utility pole. At least one power line was brought to the ground by the crash and others were drooping closer to the ground. A sheriff's deputy was first on the scene having witnessed the incident. The deputy reported the downed power lines and instructed the occupants of the vehicle to stay inside. Responding emergency personnel were advised of the downed power lines over the radio and verbally.

Captain Rhea arrived as the driver of a fire department squad. He carried equipment to other firefighters and responders that were assessing injuries to the truck's occupants and provided treatment. As he walked on the scene, Captain Rhea attempted to step over a power line on the ground, stumbled, attempted to regain his balance, and came into contact with one or more power lines. He was fatally electrocuted.

January 25, 2009–1405 hrs
Cory James Galloway, Firefighter
Age 23, Career • *Kilgore Fire Department, Texas*

Kyle Wayne Perkins, Firefighter
Age 45, Career • *Kilgore Fire Department, Texas*

The Kilgore Fire Department had recently received a new 95-foot midmount aerial ladder tower. Firefighters had received training from a manufacturer's representative. The apparatus was not yet in service.

Firefighters brought the ladder truck to a highrise dormitory building at a local college. Three firefighters were in the platform of the apparatus as it ascended to the roof. When the platform arrived at the roof of the building, the operator set the platform down on the parapet of the building.

During the maneuver, the platform became stuck on the edge of the parapet. The operator attempted to free the platform. When the platform dislodged from the parapet, it moved violently away from the building and then whipped back and forth. During this time, Firefighter Galloway and Firefighter Perkins fell from the platform to their deaths.

None of the three occupants of the platform were wearing restraints or ladder belts.

For additional information regarding this incident, please refer to NIOSH Fire Fighter Fatality Investigation and Prevention Program Report F2009-06 (http://www.cdc.gov/niosh/fire/reports/face200906.html).

January 30, 2009–1405 hrs
Charles D. Myshrall, Firefighter/EMT
Age 67, Volunteer • *North Coventry Volunteer Fire Department, Inc., Connecticut*

Firefighter/EMT Myshrall responded to a medical emergency incident with other members of his fire department. As the incident concluded, Firefighter/EMT Myshrall slipped on the driveway at the scene of the incident. Firefighter/EMT Myshrall was transported to the hospital.

Firefighter/EMT Myshrall sustained an injury to his leg that required surgery. The surgery was conducted on February 5, 2009. Firefighter/EMT Myshrall died on February 26, 2009, as he recovered from his surgery. The cause of death was a blood clot in his lungs (pulmonary embolism). The autopsy indicated that the clot was related to his leg injury.

January 30, 2009–2358 hrs
Mark Bradley Davis, Firefighter/EMT
Age 25, Volunteer • Cape Vincent Volunteer Fire Department, New York

Firefighter/EMT Davis and members of his fire department were dispatched to an emergency medical incident in a residence. As the patient was treated, he became agitated and went into a bedroom. The patient emerged from the bedroom armed with a rifle. He fired two shots at responders. One shot struck Firefighter/EMT Davis.

The gunman was placed into police custody and responders were able to begin treatment of Firefighter/EMT Davis. Despite their efforts, he was pronounced dead at a local hospital. The cause of death was listed as exsanguination (blood loss).

January 31, 2009–0040 hrs
William Gray Parsons, Engineer
Age 58, Volunteer • Millers Creek Volunteer Fire Department, North Carolina

Firefighters responded to the report of a structure fire in a residence. Upon their arrival, firefighters discovered a fire in a laundry room. The fire was knocked down when there was a report of an ill firefighter at the scene.

Engineer Parsons responded as the driver of a tanker (tender). He began to feel ill during the response and complained to another firefighter about chest pains upon his arrival. Firefighters and paramedics provided treatment at the scene and Engineer Parsons was transported to a local hospital. He was flown to a regional care facility but died on February 1, 2009. The cause of death was listed as a heart attack.

February 9, 2009–2144 hrs
Manuel "Manny" Rivera, Firefighter
Age 42, Career • Trenton Fire Department, New Jersey

Firefighter Rivera was a member of the crew of an engine company dispatched to a structure fire in a residence. When firefighters arrived on the scene, they found a working fire on the first floor of the structure with smoke showing from the second and third floors. A civilian in need of rescue was noted at one of the third floor windows.

Continued on next page.

Firefighter Rivera raised a 24-foot ground ladder to the window, climbed the ladder, and assisted the trapped civilian to the ground. In the course of communicating with the civilian, Firefighter Rivera was forced to remove his self-contained breathing apparatus (SCBA) facepiece in order to communicate.

When he arrived back on the ground, Firefighter Rivera collapsed due to a heart attack. He was treated on the scene by firefighters and emergency medical personnel and then transported to the hospital in cardiac arrest. Emergency room personnel were able to restore a heartbeat and he was placed on a ventilator. He was transferred to a regional care facility where he died on March 31, 2009.

February 9, 2009–1900 hrs
Dean Walter Mathison, Fire Chief
Age 63, Volunteer • *Clayton-Winchester Fire Department, Wisconsin*

Chief Mathison and other members of his fire department attended a regional safety organization meeting. The event was part business meeting and part training session. At the conclusion of the training, firefighters ate a light dinner. At about 2235 hours, Chief Mathison was discovered, unconscious, in the fire station where the meeting and training was held.

Chief Mathison was treated by other firefighters and transported to the hospital. Chief Mathison did not recover and was pronounced dead the next day. The cause of death was a heart attack.

February 10, 2009–Time Unknown
Jeffrey Isbell, Firefighter
Age 33, Career • *Enterprise Fire Department, Alabama*

Firefighter Isbell and another firefighter from the Enterprise Fire Department were attending a trench rescue course at the Alabama Fire College. At the conclusion of classes for the day, Firefighter Isbell and the other firefighter had dinner and then returned to their hotel room.

The next morning, Firefighter Isbell's alarm went off and he did not rise from bed. The other firefighter checked on Firefighter Isbell and found him to be unresponsive. He was treated and transported to the hospital but did not recover. The cause of death was found to be a congenital structural issue with Firefighter Isbell's heart. He had been a firefighter for 6 years and had been cleared for firefighter duties by a physician when he joined the Enterprise Fire Department in 2006.

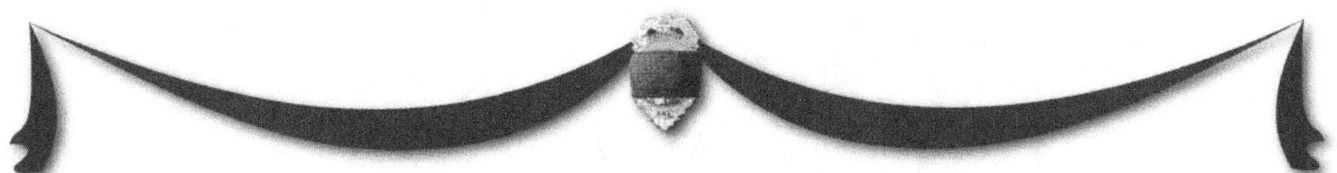

February 15, 2009–2303 hrs
Albert G. Eberle, Jr., Fire Police Captain
Age 74, Volunteer • *Roslyn Fire Company, Pennsylvania*

Fire Police Captain Eberle was working traffic control at a vehicle crash. While on scene, Fire Police Captain Eberle suffered a heart attack. Cardiopulmonary resuscitation (CPR) was performed at the scene and Fire Police Captain Eberle was transported to a local hospital where he later was pronounced dead.

February 19, 2009–2157 hrs
Johnnie Howard Hammons, Lieutenant
Age 49, Volunteer • *Craigsville-Beaver-Cottle (C-B-C) Volunteer Fire Department, West Virginia*

Timothy Allen "Little Tim" Nicholas, Firefighter
Age 26, Volunteer • *Craigsville-Beaver-Cottle (C-B-C) Volunteer Fire Department, West Virginia*

Firefighters were dispatched to a report of a camper on fire. The camper was located next to a manufactured home and flames were impinging on the home. The occupants of the camper and manufactured home were able to evacuate prior to fire department arrival.

Lieutenant Hammons and Firefighter Nicholas advanced an attack line into the home as other firefighters attacked the fire in the camper. Five to 10 minutes after their entry, the pump operator sounded an evacuation signal, concerned that he was running out of water.

When Lieutenant Hammons and Firefighter Nicholas did not emerge from the home, firefighters called out for them, attempted to contact them on the radio, and tugged on the attack line to no avail. The firefighters were eventually discovered in the front room of the home unconscious. Both firefighters were pronounced dead at the scene.

Both firefighters were equipped with full structural firefighting protective clothing and SCBA. Neither firefighter was equipped with a personal alert safety system (PASS) device. Both firefighters were found with their helmets in place, with the SCBA regulator attached to the facepiece, but the firefighters were not wearing facepieces.

The cause of death for both firefighters was listed as smoke inhalation and thermal burns. Both firefighters also had lethal doses of cyanide in their blood at autopsy.

For additional information regarding this incident, please refer to NIOSH Fire Fighter Fatality Investigation and Prevention Program Report F2009-07 (http://www.cdc.gov/niosh/fire/reports/face200907.html).

John Wayne Adams, Firefighter
Age 45, Volunteer • *Silver City Volunteer Fire Department, Oklahoma*

Firefighter Adams and the members of his fire department were dispatched to a report of three wildland fires near a ranch in their district. The incident was dispatched at 1026 hours. When Firefighter Adams arrived at the fire station, two of the department's brush trucks had already responded to the incident.

Firefighter Adams donned his wildland firefighting personal protective equipment (PPE) and then assisted another firefighter as he prepared a reserve brush truck for use. The Incident Commander (IC) ordered the unit to respond to the scene.

Firefighter Adams participated in firefighting, mopup, and fire watch duties. At approximately 2200 hours, Firefighter Adams was assigned as the driver of a brush truck. He stopped the unit and was discovered unresponsive at the wheel of the brush truck by other firefighters. CPR was initiated and an ambulance was called. Although his condition temporarily improved, Firefighter Adams did not recover and was pronounced dead by a medical helicopter crew at 2317 hours.

The death certificate and the autopsy completed by the medical examiner listed "atherosclerotic and hypertrophic cardiovascular disease" as the cause of death.

For additional information regarding this incident, please refer to NIOSH Fire Fighter Fatality Investigation and Prevention Program Report F2009-09 (http://www.cdc.gov/niosh/fire/reports/face200909.html).

Tommy Lee Adams, Battalion Chief
Age 52, Career • *Shreveport Fire Department, Louisiana*

Chief Adams was participating in the Krewe of Gemini Mardi Gras parade. At the conclusion of the parade, the Shreveport engine and ladder that participated in the parade were stopped by the roadside, the riders were removed from the apparatus, and items on the top of the trucks were being accounted for and secured. Chief Adams was on the top of the ladder truck.

Chief Adams fell from the apparatus to the pavement below. He suffered severe injuries including a brain injury. He was treated by firefighters and transported to the hospital. Chief Adams did not recover from his injuries and died on December 12, 2009.

February 23, 2009–1559 hrs
Derek Edward North, Firefighter

Age 34, Career • *Stockton Fire Department – Lanier County, Georgia*

Firefighter North was the front seat passenger in a 1964 "Fire Knocker" engine responding to a wildland fire. As the engine responded, it approached a split intersection with another highway. As the apparatus approached the intersection, the driver tried to apply the brakes and found that he was unable to stop.

The apparatus traveled through the first part of the intersection and then swerved to avoid traffic in the second part of the intersection. The apparatus overturned and came to rest on a utility pole. Firefighter North was ejected from the vehicle during the crash and sustained fatal injuries.

The cause of death was listed as trauma to the head. Neither the driver of the engine nor Firefighter North were wearing seatbelts at the time of the crash.

For additional information regarding this incident, please refer to NIOSH Fire Fighter Fatality Investigation and Prevention Program Report F2009-08 (http://www.cdc.gov/niosh/fire/reports/face200908.html).

February 27, 2009–1620 hrs
Michael James "Mike D" Darrington, Paramedic Firefighter

Age 45, Career • *Toledo Fire and Rescue Department, Ohio*

Paramedic Firefighter Darrington was on duty at his regular assignment at Engine 14. The engine responded to three incidents before 1300 hours. After returning from the third incident, Paramedic Firefighter Darrington returned to the crew's quarters on the second floor of the fire station. Earlier in the shift, Paramedic Firefighter Darrington had told another firefighter that he was not feeling well.

At 1608 hours, Engine 14 was dispatched to a structure fire in their first-due area. Paramedic Firefighter Darrington did not report to the apparatus for the response. Firefighters found him unconscious on the second floor of the station having suffered a heart attack.

Firefighters provided medical attention and an ambulance was dispatched to the station. Despite their efforts, Paramedic Firefighter Darrington was not revived and was pronounced dead by paramedics at 1620 hours.

March 1, 2009–1810 hrs
Alan Mack Hermel, Firefighter
Age 62, Volunteer • Bossier Parish Fire District 1, Louisiana

Firefighter Hermel and the members of his fire department responded to a motor vehicle crash at 1810 hours on March 1, 2009. No injuries were found and firefighters returned to quarters by 1829 hours.

At 0248 hours that night, Firefighter Hermel's spouse called 9-1-1 to report that Firefighter Hermel was ill. Firefighters responded to his residence and found Firefighter Hermel exhibiting signs of a stroke. He was treated at the scene and transported to the hospital by ambulance. His condition worsened and Firefighter Hermel died as the result of an intercranial bleed on March 3, 2009.

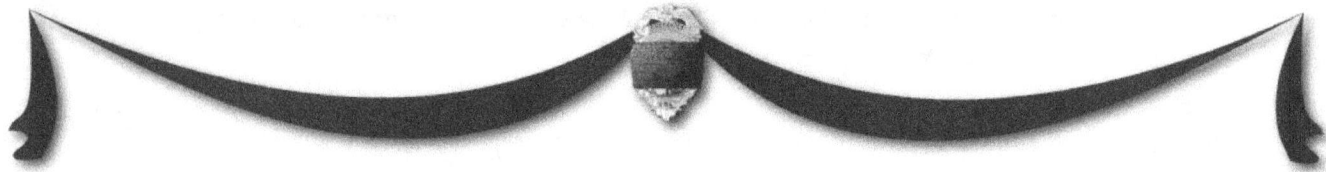

March 4, 2009–1428 hrs
Christopher Allen Dill, Corporal
Age 43 – Career • Oklahoma City Fire Department, Oklahoma

Corporal Dill and the members of his engine company were dispatched on the second alarm to an apartment building fire at 1428 hours. Upon their arrival, Corporal Dill's engine was assigned as the Rapid Intervention Team (RIT).

After the fire was knocked down, Corporal Dill and his crew were assigned to do overhaul on the structure. After approximately 20 minutes of work, the crew rested and drank water. When their overhaul duties were completed, Corporal Dill and his crew assisted other crews in picking up tools and hoselines.

As the crew prepared to leave the scene, Corporal Dill told a member of his crew that his back was hurting. Corporal Dill sat down on the ground, began to feel worse, and asked for assistance. Corporal Dill lost consciousness and collapsed. Firefighters and responding EMS workers provided assistance and Corporal Dill was transported to the hospital. Despite efforts on the scene, in the ambulance, and at the hospital, Corporal Dill was pronounced dead at 1643 hours, 43 minutes after his collapse. His death was caused by a heart attack.

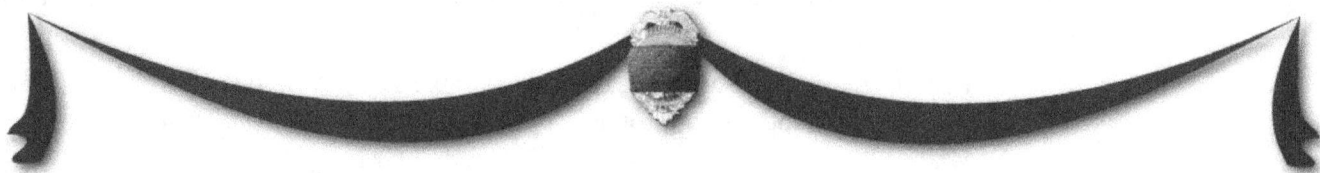

March 11, 2009–Time Unknown
Gregory Carroll Cooke, Firefighter
Age 60, Volunteer • Salem Volunteer Fire Department, North Carolina

On March 11, 2009, Firefighter Cooke was working at a woods fire when he went into cardiac arrest. He was transported by air to the Wake Medical Center in Raleigh, NC. On March 21, 2009, Firefighter Cooke passed away.

March 14, 2009–1327 hrs
William Roger Vorwark, Lieutenant
Age 49, Career • *Odessa Fire and Rescue Protection District, Missouri*

Lieutenant Vorwark was a career firefighter for the Odessa Fire and Rescue Protection District. He was also permitted to respond as a paid-on-call firefighter when he was off duty.

On a day off, he responded to two wildland fires. At the second incident, firefighters were unable to access the fire with brush trucks. Lieutenant Vorwark wore full structural protective clothing, without an SCBA, and used a portable water backpack to fight the fire for nearly 40 minutes.

As Lieutenant Vorwark and other firefighters rested, Lieutenant Vorwark suddenly collapsed. CPR was initiated and a ground ambulance and medical helicopter were called to the scene. Lieutenant Vorwark was carried by a tractor to a paved roadway. An automated external defibrillator (AED) was attached and delivered a shock with no change in condition. Paramedics arrived and continued to treat Lieutenant Vorwark. He was flown by medical helicopter to a local hospital where treatment was continued until he was pronounced dead over an hour after his collapse. The cause of death was listed as a heart attack.

For additional information regarding this incident, please refer to NIOSH Fire Fighter Fatality Investigation and Prevention Program Report F2010-01 (http://www.cdc.gov/niosh/fire/reports/face201001.html).

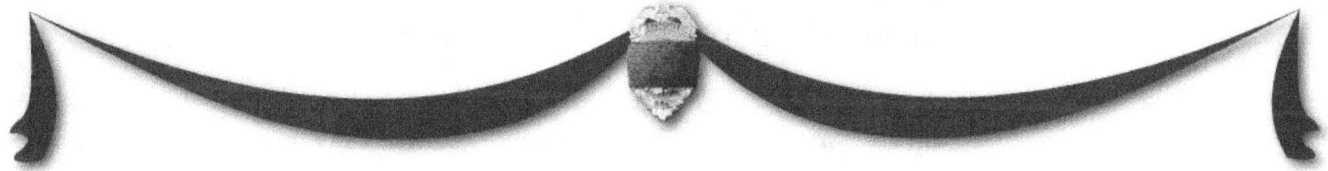

March 23, 2009–1150 hrs
Nolan Ray Schmidt, Fire Chief
Age 37, Volunteer • *Hydro Volunteer Fire Department, Oklahoma*

Chief Schmidt and the members of his fire department were dispatched to a report of a possible fire in a large grain bin. Firefighters entered the bin to investigate. Chief Schmidt ordered firefighters to exit the bin. In order to get out of the bin, firefighters had to climb up a long ladder. One of the firefighters in the bin was fatigued and could not complete the climb. Chief Schmidt entered the bin the assist the firefighter.

Both firefighters subsequently lost consciousness. Firefighters on the exterior cut a hole in the metal wall of the bin and extricated the two firefighters. Chief Schmidt was transported to the hospital but was pronounced dead.

The cause of death was listed as asphyxiation due to probable carbon monoxide toxicity. At autopsy, the carboxyhemoglobin level in Chief Schmidt's blood was found to be 58 percent.

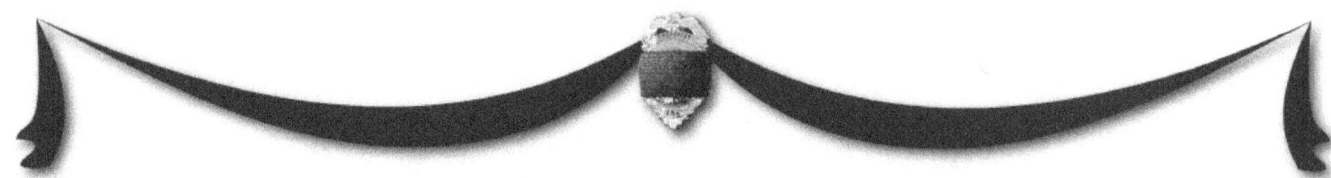

March 26, 2009–2145 hrs
Robert L. Strang, Lieutenant
Age 60, Career • *Melbourne Fire Department, Florida*

Lieutenant Strang and the members of his crew responded to a structure fire. Lieutenant Strang's engine company was first on the scene and advanced an attack line into the structure. The fire was extinguished and units returned to quarters.

When they arrived back at quarters, Lieutenant Strang retired to his bunkroom. Approximately 34 minutes after returning to quarters, Lieutenant Strang was discovered unconscious. Firefighters removed him from his room and initiated advanced life support care. Lieutenant Strang was transported to the hospital where he was pronounced dead. The cause of death was a heart attack.

March 27, 2009–0630 hrs
Michael Martin Gilbreath, Fire Chief
Age 55, Volunteer • *Double Springs Fire Department, Alabama*

Chief Gilbreath and the members of his fire department were dispatched to a "trees down" call on a local county road. As he cleared the roadway, Chief Gilbreath was struck by a passing vehicle. The cause of death was listed as blunt force trauma.

March 30, 2009–2030 hrs
Robert Andrew France, Firefighter/EMT – Training Officer
Age 45, Volunteer • *Stateline Volunteer Fire Department, Mississippi*

Firefighter France attended a fire training meeting at the Green County Emergency Services Office. Upon completion of the meeting, he was returning to his home jurisdiction on his motorcycle when he struck a deer. Firefighter France was treated on scene by emergency services personnel and subsequently transported via helicopter to the USA Medical Center in Mobile, AL, where he succumbed to his injuries.

April 1, 2009–2126 hrs
George Albert Wimberly, Firefighter
Age 63, Volunteer • *Stonewall Volunteer Fire Department, Mississippi*

Firefighter Wimberly and the members of his fire department were paged to respond to a residential structure fire. Firefighter Wimberly was responding in his personal vehicle when he began to experience severe chest pains.

Firefighter Wimberly pulled to the roadside approximately one-half mile from the fire scene. He was transported to the hospital where he was pronounced dead due to a heart attack.

April 1, 2009–Time Unknown
John William Jeffers, Assistant Chief
Age 54, Volunteer • *Wellington-Greer Fire Protection District, Illinois*

Following a meeting at the fire station, Assistant Chief Jeffers and other firefighters were testing fire hoses. After reloading a hose on an engine, Assistant Chief Jeffers sat down on the tail board of the apparatus, and told other firefighters that he was not feeling well, and asked to be taken home. Assistant Chief Jeffers described his illness as the flu and told others that he had vomited.

A friend stopped by the next afternoon to see him and when there was no response at the door, the fire department was called. After gaining access to the home they discovered Assistant Chief Jeffers had died from a heart attack.

April 4, 2009–2118 hrs
John DeWayne Weber, Firefighter
Age 77, Volunteer • *Township Fire Department, Wisconsin*

Firefighter Weber and the members of his fire department were dispatched to a report of a wildland fire. Firefighter Weber lived next door to the fire station. He rode an all terrain vehicle to the fire station and began to prepare the station's four fire trucks for their response. This duty consisted of starting the apparatus and pulling them out onto the apron.

Another firefighter arrived and responded to the incident in a tanker (tender). Firefighter Weber remained behind in case another apparatus was needed. Fire department units from other fire stations also responded to the incident.

At approximately 2230 hours, Firefighter Weber was discovered unconscious in the apparatus bay. CPR was initiated and an ambulance responded. Firefighter Weber was evaluated by paramedics and further resuscitative measures were discontinued. He had died of a heart attack.

April 8, 2009–1426 hrs
Heath Jeffrey Van Handel, Pilot

Age 36, Wildland Full-Time • *Wisconsin Department of Natural Resources, Wisconsin*

Pilot Van Handel was conducting aerial reconnaissance of a wildland fire in Wood County in a Cessna 337 aircraft. He circled the fire area three times and then the aircraft crashed. Pilot Van Handel was killed on impact.

A National Transportation Safety Board (NTSB) investigation of the crash indicated that the probable cause was a failure to maintain adequate airspeed which resulted in an aerodynamic stall at a low altitude.

For additional information about this crash, consult the NTSB website at http://www.ntsb.gov/ntsb/query.asp - NTSB identification CEN09FA242.

April 11, 2009–1100 hrs
Patrick J. Reardon, Firefighter

Age 42, Career • *New Haven Fire Department, Connecticut*

Firefighter Reardon had just returned from an incident response and was back in his fire station. He sustained a fall on the apparatus floor for an unknown reason and suffered severe head injuries. Firefighter Reardon was transported to the hospital and died on April 21, 2009, as a result of his injuries.

April 12, 2009–0014 hrs
James Arthur Harlow, Sr., Captain

Age 49, Career • *Houston Fire Department, Texas*

Damion Jon Hobbs, Probationary Firefighter

Age 30, Career • *Houston Fire Department, Texas*

Captain Harlow, Firefighter Hobbs, and the members of Engine 26 were dispatched to the report of a structure fire in a residence just after midnight. Engine 26 was the first to arrive on the scene approximately 6 minutes after dispatch. They found a large residence with smoke showing.

Captain Harlow and Firefighter Hobbs advanced an attack line to the interior of the structure. Firefighter Hobbs was on the nozzle, backed up by Captain Harlow and another firefighter. As the line was advanced, the third firefighter had to withdraw from the structure due to an issue with his protective clothing.

Continued on next page.

Approximately 7 minutes after the arrival of Engine 26 on the scene, fire conditions changed rapidly and firefighters exited the building. The IC declared a defensive strategy and apparatus air horns were sounded to announce the evacuation of the building. Firefighters realized that Captain Harlow and Firefighter Hobbs were not accounted for.

Firefighters used master streams and handlines to try to control the fire. Approximately 30 minutes after the switch to defensive strategy, firefighters were able to enter the structure and Captain Harlow and Firefighter Hobbs were found.

Both firefighters were removed to the front yard of the home and resuscitative efforts were made, to no avail. The cause of death for Captain Harlow was listed as thermal injuries and smoke inhalation. The cause of death for Firefighter Hobbs was listed as thermal injuries.

This incident was the first structure fire response for Firefighter Hobbs.

For additional information regarding this incident, please refer to NIOSH Fire Fighter Fatality Investigation and Prevention Program Report F2009-11 (http://www.cdc.gov/niosh/fire/reports/face200911.html).

The Texas State Fire Marshal's Office prepared a detailed report on this incident. The report is available at http://www.tdi.state.tx.us/fire/fmloddinvesti.html

April 15, 2009–1945 hrs
Charles Fletcher "Buck" Clough, Jr., Fire Chief
Age 41, Volunteer
Sudlersville Fire Company, Inc., Maryland

Chief Clough and the members of his fire department were dispatched to the report of an appliance fire. Chief Clough responded from his residence in a fire department pickup truck.

As he responded, Chief Clough came upon a car that pulled out into his path. Chief Clough attempted to maneuver around the car but his vehicle turned sideways and began to skid. His vehicle then went off the roadway and struck a tree. At approximately 1951 hours, the dispatch center received reports of a vehicle crash involving a fire department pickup truck.

Arriving firefighters discovered Chief Clough entrapped after a collision with a tree. He was pronounced dead at the scene. The cause of death was listed as multiple trauma. Chief Clough was wearing his seatbelt at the time of the crash.

For additional information regarding this incident, please refer to NIOSH Fire Fighter Fatality Investigation and Prevention Program Report F2009-12 (http://www.cdc.gov/niosh/fire/reports/face200912.html).

April 17, 2009–1900 hrs
Stephen Michael Cospelich, Lieutenant
Age 57, Career • Philadelphia Fire Department, Pennsylvania

Lieutenant Cospelich and the members of his ladder company assisted with operations at a structure fire that concluded at approximately 1500 hours. Lieutenant Cospelich went off duty at approximately 1800 hours.

While walking his dog at home, Lieutenant Cospelich suffered a stroke. He was transported from the scene but died as the result of his injury on May 19, 2009.

April 21, 2009–1800 hrs
Dennis M. Simmons, District Chief
Age 63, Volunteer • Stafford County Fire/EMS Department, Kansas

A number of different fire department units responded to assist with fighting a wildland fire that resulted from a controlled burn. District Chief Simmons assisted with firefighting efforts from his arrival on the scene at approximately 1544 hours until approximately 1715 hours. At that time, District Chief Simmons told the IC that he was exhausted and needed to be relieved.

District Chief Simmons departed the scene in his personal vehicle. At approximately 1800 hours, reports were received of a vehicle in a ditch. Responders discovered District Chief Simmons. He was unresponsive. He was transported to the hospital but died some time later. The cause of death was a heart attack.

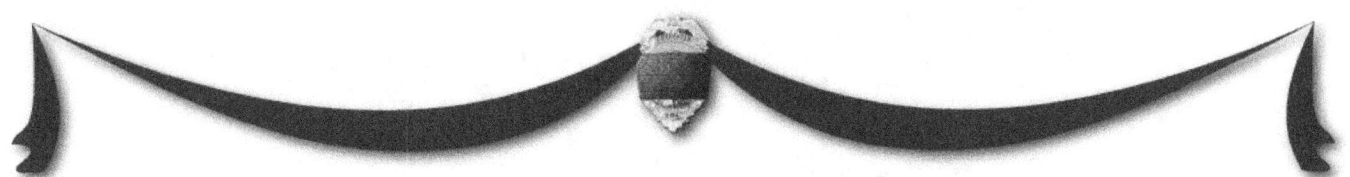

April 25, 2009–1000 hrs
Thomas Lowell Risk, Captain/Pilot

Age 66, Wildland Contract • Neptune *Aviation Services under contract to the* U.S. *Forest Service*

Michael Wayne Flynn, First Officer/Copilot

Age 59, Wildland Contract • Neptune *Aviation Services under contract to the* U.S. *Forest Service*

Brian Joseph Buss, Crew Chief

Age 32, Wildland Contract • Neptune *Aviation Services under contract to the* U.S. *Forest Service*

Neptune Aviation Services received a resource order to relocate Tanker 42 from its base in Missoula, MT, to Alamogordo, NM, to assist with wildland firefighting operations. Tanker 42 was a 1962 Lockheed P2V-7 Neptune aircraft. The aircraft departed Missoula at 0803 hours.

According to the NTSB report, the aircraft appeared to be negotiating around some weather systems as it flew over Utah. The aircraft crashed into a slope at an elevation of approximately 5,600 feet. All three firefighters were killed in the crash.

For additional information about this crash, consult the NTSB website at http://www.ntsb.gov/ntsb/query.asp - NTSB identification WPR09GA216.

April 29, 2009–0710 hrs
Cohnway Matthew Johnson, Cadet Firefighter

Age 26, Career • Houston *Fire Department,* Texas

Cadet Firefighter Johnson was nearing the end of a 4-mile cadence run. He was about 60 yards from the finish when he began to stagger. He was helped to the ground and attended to by a paramedic. He was transported to the hospital. At the hospital, his body temperature was found to be 105.3 °F.

Firefighter Johnson died on May 4, 2009. The cause of death was listed as complications of hyperthermia and dehydration.

May 10, 2009–2200 hrs
Frankie Paul Nelson, Fire Captain
Age 51, Career • Shreveport Fire Department, Louisiana

Captain Nelson worked a 24-hour shift that began at 1400 hours on May 9, 2009, and ended at 1400 hours on May 10, 2009. During the shift, Captain Nelson responded to a motor vehicle crash and provided Incident Command, EMS assistance, and traffic safety and control measures.

Approximately 8 hours after going off duty, Captain Nelson suffered a fatal heart attack.

May 22, 2009–1523 hrs
Paul James Roberts, Firefighter
Age 54, Career • Beverly Fire Department, Massachusetts

Firefighter Roberts reported for duty at 0700 hours. He and his crew responded to two emergency medical incidents during the morning. One incident was a significant motor vehicle crash that involved the treatment of several patients.

At 1520 hours, another Beverly engine was dispatched to a 3rd alarm fire. Firefighter Roberts' engine was to cover the vacant fire station and be prepared to respond. Firefighter Roberts did not report to the apparatus for the response. He was discovered unconscious in his dorm room.

Firefighters provided treatment and Firefighter Roberts was transported to the hospital by ambulance. He was pronounced dead at the hospital. His death was caused by a heart attack.

May 30, 2009–1633 hrs
James M. "Marty" Hall, Firefighter/EMT
Age 50, Volunteer • Greentown Volunteer Fire Department, Ohio

Firefighter/EMT Hall was on duty with the Greentown Fire Department on May 29, 2009, from 0600 to 1800 hours. During that time, he responded to two EMS incidents. On May 30, 2009, while participating in a charity basketball game for his employer, the Canton Fire Department, he collapsed due to a medical emergency. Care was initiated on scene and he was transported to the local medical facility where, despite resuscitative efforts, he succumbed to his illness (heart attack).

Firefighter/EMT Hall was also a career Captain with the Canton Fire Department.

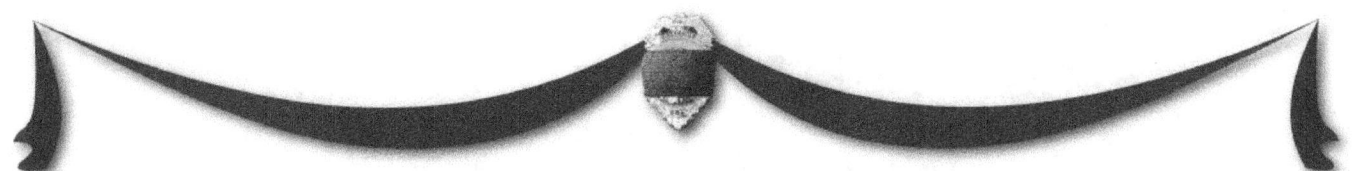

June 4, 2009–2156 hrs
Jeffrey Houston Reed, Firefighter
Age 39, Career • *Pulaski Fire Department, Virginia*

Firefighter Reed and other firefighters responded to the scene of a structure fire in an abandoned building. Firefighter Reed operated the pump on an engine. After about 3 hours of onscene operations, Firefighter Reed complained of a headache and told others that he could not see out of one eye.

He was transported to the hospital. Tests at the hospital revealed a colloid cyst in his brain. Firefighter Reed died as a result of his illness on June 8, 2009.

For additional information regarding this incident, please refer to NIOSH Fire Fighter Fatality Investigation and Prevention Program Report F2009-22 (http://www.cdc.gov/niosh/fire/reports/face200922.html).

June 5, 2009–0053 hrs
Matthew Douglas Tramel, Firefighter
Age 18, Volunteer • *Town of Pembroke Fire Department, North Carolina*

Firefighter Tramel was responding to a car fire in his personal vehicle when he lost control of the vehicle, crossed the center line, left the roadway, and struck a tree. Firefighter Tramel was killed as a result of massive head trauma sustained during the crash. He was not wearing a seatbelt at the time of the crash. Speed and wet road conditions were cited as factors in the crash.

June 10, 2009–0630 hrs
Debra Ann Cole, Firefighter/Paramedic
Age 40, Career • *South Portland Fire Department, Maine*

Firefighter/Paramedic Cole collapsed in the station while on duty and after having run at least one emergency call during the shift. Her collapse was not witnessed but was overheard and she was tended to immediately by the rest of her crew. Firefighter/Paramedic Cole was transported to Maine Medical Center in Portland, where she was treated and underwent almost 11 hours of surgery. Firefighter/Paramedic Cole succumbed to her injury at approximately 1630 hours the following day. The cause of death was listed as a stroke.

June 10, 2009–Time Unknown
Conrad A. Mansfield, Firefighter

Age 45, Volunteer • *Delaware Township Volunteer Fire Department, Ohio*

Firefighter Mansfield collapsed while participating in pump training. Mansfield was transported to a local hospital and then airlifted to a hospital in Toledo where he was placed on life support. Firefighter Mansfield succumbed to his injury on June 12, 2009. Preliminary results indicated the nature of the fatal injury was a stroke, but a final determination is pending further investigation.

June 16, 2009–1800 hrs
Lyle Chester Lewis, Firefighter

Age 50, Paid-on-Call • *Osborne County Rural Fire District #3, Kansas*

Firefighter Lewis and the members of his fire department responded to a residential structure fire. It was a hot and humid day. Firefighters found a working fire.

Firefighter Lewis and other firefighters advanced a handline through a second story window and operated in the interior. After exiting the structure temporarily, Firefighter Lewis and another firefighter went back into the structure to continue operations. As Firefighter Lewis exited the building for the second time, his low air alarm was sounding.

When he arrived at rehab, he told other firefighters that he was overheated. He was assessed and found to be suffering from chest pains. He was transported to the hospital by ambulance. Firefighter Lewis lost consciousness; an AED was attached and delivered a shock.

Treatment efforts continued in the emergency room but were not successful. Firefighter Lewis was pronounced dead at 1845 hours.

For additional information regarding this incident, please refer to NIOSH Fire Fighter Fatality Investigation and Prevention Program Report F2010-02 (http://www.cdc.gov/niosh/fire/reports/face201002.html).

June 16, 2009–1845 hrs
James E. "Jimmy" Cameron, Firefighter

Age 47, Volunteer • *South Chester Fire Department, South Carolina*

Firefighter Cameron and the members of his fire department were dispatched to perform a storm damage assessment after a serious storm passed through the area. While performing the assessment, Firefighter Cameron suffered a heart attack and died in his vehicle at the side of the road. He was found by other firefighters.

June 18, 2009–0715 hrs
William V. "Bill" Thompson, Sr., Firefighter
Age 66, Volunteer • Dushore Fire Company #1, Pennsylvania

Firefighter Thompson suffered an apparent heart attack and died at approximately 0715 hours on June 18, 2009, while on his way to work. Firefighter Thompson had responded to an emergency call the previous evening.

June 26, 2009–1630 hrs
Brett Michael Stearns, Captain
Age 29, Wildland Full-Time • Bureau of Land Management – Little Snake Field Office
Northwest Colorado Fire Management Unit

Captain Stearns and a dozen other firefighters were conducting a hazardous tree abatement training project near Craig, Colorado. During the operation, a tree fell and struck Captain Stearns in the head and back.

Captain Stearns was treated at the scene and eventually pronounced dead by an emergency room physician through radio communications.

A comprehensive report on the incident was prepared by the Northwest Colorado Fire Management Unit. It can be found at http://www.blm.gov/pgdata/etc/medialib/blm/nifc/Safety.Par.41629.File.dat/FreemanFactual.pdf

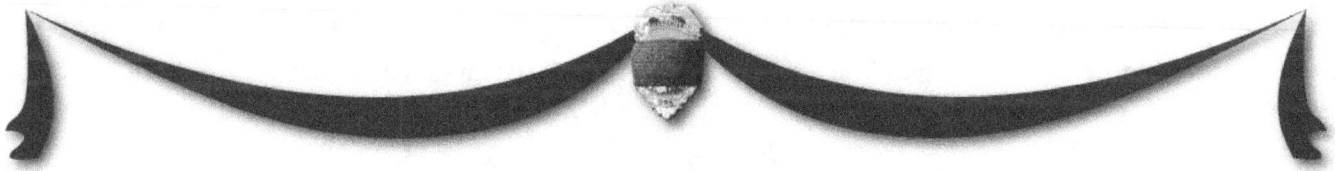

June 29, 2009–1435 hrs
Allan "Pickles" LePage, Assistant Chief
Age 67, Volunteer • Kingston Fire District, Rhode Island

Assistant Chief LePage positioned the department's tower ladder truck at the rear of the fire station. He set the stabilizers and entered the ladder's platform area. Assistant Chief LePage maneuvered the platform into the open rear apparatus bay door of the fire station. His head became trapped between the door frame and the platform railing and he received fatal injuries.

Other firefighters found him trapped, repositioned the ladder, removed Assistant Chief LePage, and provided medical treatment to no avail.

For additional information regarding this incident, please refer to NIOSH Fire Fighter Fatality Investigation and Prevention Program Report F2009-18 (http://www.cdc.gov/niosh/fire/reports/face200918.html).

July 4, 2009–2257 hrs
Dale Elliott Haddix, Assistant Fire Chief
Age 70, Volunteer • Schell City Volunteer Fire Department, Missouri

Upon arrival at the scene of a vacant residential structure fire, Assistant Fire Chief Haddix told other firefighters that he would run the pump on the attack engine from inside the cab of the apparatus. A short time later, he was observed slumped over the wheel. Assistant Chief Haddix was treated on scene by a standby medical crew but succumbed to his injury.

An 18-year-old man was charged with arson for shooting a roman candle into the structure and causing the structure fire.

July 6, 2009–0354 hrs
Ryan Wingard, Firefighter
Age 28, Volunteer • Strattanville Volunteer Fire Company, No. 1, Pennsylvania

Firefighter Wingard and the members of his fire department responded to a debris fire on the site of a house that was recently demolished. Firefighter Wingard assisted with hoseline deployment and then told other firefighters that he was ill.

Firefighter Wingard collapsed and CPR was initiated. An ambulance arrived on the scene and transported Firefighter Wingard to the hospital. He was pronounced dead at the hospital. The cause of death was a cardiac condition.

For additional information regarding this incident, please refer to NIOSH Fire Fighter Fatality Investigation and Prevention Program Report F2009-24 (http://www.cdc.gov/niosh/fire/reports/face200924.html).

July 7, 2009–1545 hrs
David V. Grass, Jr., Firefighter
Age 34, Volunteer • Ste. Genevieve Fire Department, Missouri

Firefighter Grass was participating in physical fitness activities at his fire station. As he was exercising, he felt a pop in his head and began to feel ill. He called his wife to the fire station in order to go to the hospital. Firefighter Grass's wife transported him to a local hospital.

Firefighter Grass was transferred by medical helicopter to a regional hospital. He was treated at the hospital but his condition did not improve. Firefighter Grass died as the result of a stroke on July 8, 2009.

July 9, 2009–0852 hrs
Joseph T. Grace, Fire Equipment Operator
Age 47, Career • *Saint Tammany Fire Protection District #4 – Mandeville Fire Department, Louisiana*

Fire Equipment Operator Grace worked a 24-hour shift from 0700 hours on July 8, 2009, through 0700 hours on July 9, 2009. During the shift, he responded to two medical emergency incidents and participated in a structural firefighting training exercise in full structural protective clothing.

Fire Equipment Operator Grace went off duty and traveled to his second job as a paramedic for an ambulance service. At approximately 0852 hours, Fire Equipment Operator Grace suffered a stroke. He was transported to the hospital but did not recover. He died on July 10, 2009.

July 21, 2009–1010 hrs
Thomas David "TJ" Marovich, Jr., Firefighter
Age 20 – Wildland Full-Time • *USDA Forest Service – Modoc National Forest, California*

Firefighter Marovich was participating in required proficiency training that involved rappelling from a helicopter. Firefighter Marovich fell approximately 200 feet to the ground. It appeared that Firefighter Marovich was not properly attached to the rope, resulting in a free fall. Firefighter Marovich died of multiple blunt trauma injuries.

July 31, 2009–Time Unknown
Eric Allen Tinkham, Captain/Paramedic
Age 44, Career • *Queen Creek Fire Department, Arizona*

Captain Tinkham reported for a 24-hour shift on July 31, 2009. During the shift, he and his crew responded to an emergency medical incident and conducted training in full structural protective clothing, including SCBA. Captain Tinkham retired for the evening at approximately 2230hrs.

When he did not enter the kitchen of the fire station on the morning of August 1, 2009, a member of his crew went to check on Captain Tinkham. He was found unconscious and unresponsive.

Captain Tinkham was treated by firefighters and transported to the hospital. Despite these efforts, he was pronounced dead. The cause of death was a heart attack.

August 6, 2009–1430 hrs
John P. "Jack" Horton, Assistant Chief
Age 68, Volunteer • *Marlboro Volunteer Fire Company, Inc., Vermont*

Assistant Chief Horton was in command of fire department operations at a single-vehicle rollover crash. When an ambulance arrived on the scene, Chief Horton and another firefighter decided to reposition an engine to give the ambulance better access to the scene.

As the apparatus was backed up, Chief Horton likely experienced a medical emergency that caused him to collapse in the path of the engine. The engine backed over Chief Horton and he received serious traumatic injuries. Chief Horton was transported to the hospital but did not survive.

August 12, 2009–1530 hrs
Paul V. Warhola, Firefighter
Age 47, Career • *Fire Department City of New York, New York*

Firefighter Warhola was the chauffeur of an engine company that responded to a fire alarm activation in an apartment building. As his crew investigated the alarm, Firefighter Warhola remained with his apparatus.

When firefighters emerged from the building, Firefighter Warhola told other firefighters that he was experiencing weakness, dizziness, and that he was having difficulty breathing. Firefighters transported Firefighter Warhola back to the firehouse and requested EMS. Firefighter Warhola was transported to the hospital but later died as the result of a stroke.

August 15, 2009–2006 hrs
Jimmie L. Zeeks, Fire Chief
Age 54, Volunteer • *Marion Township Rural Fire Department, Indiana*

Fire Chief Zeeks was in command of the response to a vehicle rollover crash. Chief Zeeks suffered an apparent heart attack. Members of his department immediately started life saving efforts. Chief Zeeks was rushed to a local hospital where he succumbed to his illness.

August 20, 2009–1545 hrs
David McKay Jamsa, Pilot
Age 44, Wildland Contract • *Minuteman Aerial Application under contract to Bureau of Land Management*

Pilot Jamsa and his single engine air tanker aircraft, an Airtractor AT-802A, were assigned to the Battle Mountain Airport in Nevada. He was working on aerial firefighting operations on the Hoyt Fire.

Pilot Jamsa was in the process of conducting a fire retardant drop in the saddle area between two ridges. The aircraft did not drop retardant as planned and made contact with the ground uphill to a crest. After reaching the crest, the aircraft became airborne again for a short time before crashing. Pilot Jamsa was transported to the hospital but died as the result of injuries received in the crash.

For additional information about this crash, consult the NTSB website at http://www.ntsb.gov/ntsb/query.asp - NTSB identification WPR09GA407.

August 24, 2009–0349 hrs
Charles W. "Chip" McCarthy, Jr., Lieutenant
Age 45, Career • *Buffalo Fire Department, New York*

Jonathan S. "Sim" Croom, Firefighter
Age 34, Career • *Buffalo Fire Department, New York*

Buffalo Fire Department units were dispatched to a report of a structure fire in a convenience store. Firefighters found a working fire on their arrival and civilian reports of people trapped in the building. Both the exterior and the interior of the building were difficult to access due to numerous security measures.

Firefighters searched the building and found no occupants. Access to the building's basement could not be achieved due to fire conditions and a reinforced door. An order to evacuate the building and a switch to defensive firefighting operations was declared by the IC.

At approximately 0422 hours, Members of Rescue 1, including Lieutenant McCarthy, entered the retail area of the store to verify that all firefighters had evacuated. Less than 2 minutes after their entry, the floor collapsed and Lieutenant McCarthy fell into the basement. Lieutenant McCarthy immediately began to call for help over his portable radio. The other members of his crew were unaware of what had happened and exited the structure.

Firefighter Croom was a member of the assigned Firefighter Assist and Search Team (FAST). Firefighter Croom entered the structure, it is thought that he directly heard the calls for help from Lieutenant McCarthy. Firefighter Croom also fell into the basement in close proximity to Lieutenant McCarthy.

Continued on next page.

Fire conditions worsened and firefighters were not able to access the interior. Firefighter Croom's status was not known for some time due to errors in accountability procedures. Firefighters were able to enter the basement and located the deceased firefighters at 0918 hours.

The cause of death for both firefighters was smoke inhalation. The blood carboxyhemoglobin level at autopsy for both firefighters was above 50 percent.

For additional information regarding this incident, please refer to NIOSH Fire Fighter Fatality Investigation and Prevention Program Report F2009-23 (http://www.cdc.gov/niosh/fire/reports/face200923.html).

August 30, 2009–1311 hrs
Kenneth Eugene Frizzell, Jr., Firefighter
Age 55, Volunteer • *Charleston Volunteer Fire Department, Vermont*

Firefighter Frizzell was the lone occupant of a tanker/pumper returning to the fire station from a response. Firefighter Frizzell lost control of the apparatus while descending a hill. The apparatus left the roadway, struck a telephone pole, and rolled over. Firefighter Frizzell was ejected from the vehicle during the crash.

There were reports that Firefighter Frizzell suffered a medical emergency prior to the crash that incapacitated him. No mechanical problems were discovered when the apparatus was inspected after the crash.

August 30, 2009–1430 hrs
Tedmund D. "Ted" Hall, Captain
Age 47, Career • *Los Angeles County Fire Department, California*

Arnaldo "Arnie" Quinones, Firefighter Specialist
Age 34, Career • *Los Angeles County Fire Department, California*

Captain Hall and Firefighter Specialist Quinones were working at Los Angeles County Fire Department Mount Gleason Fire Camp 16 along with California Department of Corrections and Rehabilitation inmate firefighters. The Station Fire was ongoing in the area and there was a plan to defend in place if the fire threatened the camp.

A planned firing operation was conducted on a slope below the camp. Firefighters at the camp were sheltered in the dining hall and firefighters were positioned for structure protection. The fire behaved in a manner that was not predicted and attacked the camp. Flame fronts of over 200 feet reached the camp. Firefighters were ordered to abandon the dining hall, which caught fire, and move to vehicles. These vehicles were moved to an area of the camp where the fire had already passed. An accountability report revealed that Captain Hall and Firefighter Specialist Quinones were not accounted for.

Captain Hall and Firefighter Specialist Quinones had been assigned as the firing team. Their vehicle was discovered approximately 800 feet off of the road in a ravine. Both were deceased. The autopsies of both firefighters listed the cause of death as multiple trauma.

Continued on next page.

The Station Fire was determined to be incendiary.

The factual report prepared by the Los Angeles County Fire Department related to the Camp 16 Incident can be found at http://www.fire.lacounty.gov/top_story_images/Camp16SAIR.pdf

September 9, 2009–0302 hrs
Richard William Holst, Fire Police Captain/Chaplain
Age 60, Volunteer • Huntington Manor Fire Department, New York

Fire Police Captain Holst was the first fire department member on the scene of a fire reported in a local restaurant. He met the first chief officers arriving on the scene and helped do the building walkaround to locate the fire.

About 10 minutes after the arrival of the first chief officer, Fire Police Captain Holst collapsed. Firefighters began CPR and he was transported to the hospital. Fire Police Captain Holst did not recover. The cause of death was listed as a heart attack.

The fire was determined to be incendiary.

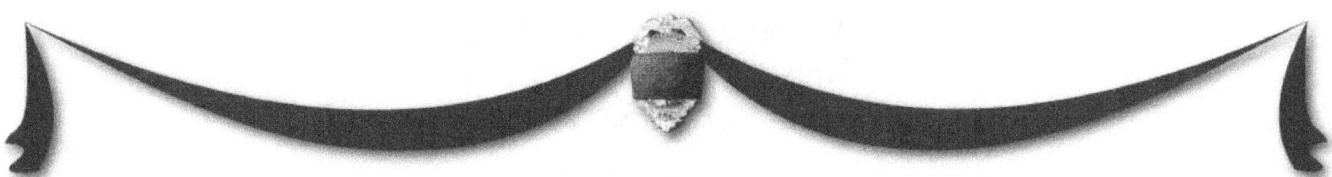

September 11, 2009–2136 hrs
Ricky R. Christiana, Firefighter/Operator
Age 44, Career • David Crockett Volunteer Fire Company #2, Louisiana

Firefighter/Operator Christiana and the members of his fire department responded to a fire in a two-story residence. When firefighters arrived, they found a working fire. Firefighter/Operator Christiana operated the pump on an engine at the scene and assisted with the deployment of attack and supply lines.

Approximately 30 minutes into operations, Firefighter/Operator Christiana collapsed. His collapse was witnessed and firefighters and paramedics came immediately to his aid. He was transported to the hospital by ambulance but did not recover. The cause of death was listed as a heart attack.

For additional information regarding this incident, please refer to NIOSH Fire Fighter Fatality Investigation and Prevention Program Report F2009-26 (http://www.cdc.gov/niosh/fire/reports/face200926.html).

September 13, 2009–0033 hrs
Terry Lynn Sharon, Firefighter
Age 60, Volunteer • Monterey Fire Department, Kentucky

Firefighter Sharon, as the driver and sole occupant of an engine, was returning to quarters from the scene of a fatal car crash. As he neared the fire station, Firefighter Sharon suffered a heart attack. Firefighter Sharon was transported to the hospital but did not survive.

The apparatus left the roadway, traveled approximately 150 feet, and ran into a camper. Damage to the apparatus and camper were minor.

October 2, 2009–0100 hrs
Patrick Joyce, Jr., Firefighter
Age 39, Career • Yonkers Fire Department, New York

Firefighter Joyce and other firefighters were operating at a working fire in a three-story residence. Firefighter Joyce and his crew were on the top floor of the structure when fire conditions changed rapidly.

Firefighter Joyce and two other firefighters were forced to jump from the structure to the ground. Firefighter Joyce received fatal injuries in the fall. The fire was incendiary.

October 3, 2009–0815 hrs
Carl M. Nordwall, Captain
Age 56, Career • Norfolk Airport Fire Department, Virginia

Captain Nordwall began complaining of chest and arm discomfort immediately following morning roll call. Firefighters began to assess his illness and Captain Nordwall collapsed. He was transported to the closest hospital as CPR was performed. Captain Nordwell did not recover and died on October 5, 2009, as the result of a pulmonary embolism.

October 17, 2009–1345 hrs
Gary David Street, Firefighter Emergency Medical Responder
Age 60, Volunteer • *East Lake Sinclair, Hancock County Station #2, Georgia*

Firefighter Street and other members of his fire department responded to a fall injury. Firefighter Street was the first responder on the scene and conducted the initial patient assessment. Firefighter Street was joined by other responders who assisted with treatment. Anticipating the arrival of the ambulance, Firefighter Street moved the department's MED truck out of the way. After completing this task, Firefighter Street exited the vehicle and collapsed.

He was treated at the scene by other responders but did not survive. The cause of death was a heart attack.

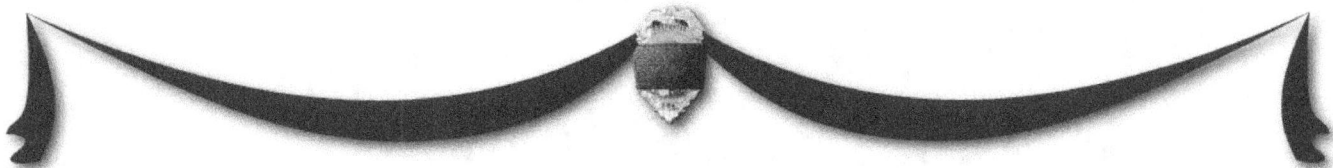

October 21, 2009–1719 hrs
John D. Thurman, Captain
Age 54, Career • *Clinton Fire Department, Mississippi*

Captain Thurman was participating in physical fitness activities in a park adjacent to his fire station. The fire department received reports that a firefighter was down in the park at 1719 hours. Responding firefighters found Captain Thurman unconscious. He was treated at the scene and transported to the hospital but did not survive. His death was caused by a heart attack.

October 24, 2009–0038 hrs
Roy Everett Westover, Jr., Lieutenant
Age 41, Volunteer • *Westover Area Volunteer Fire Company, Pennsylvania*

Lieutenant Westover responded with other members of his fire department to a working fire in a residence. Lieutenant Westover participated in fire attack operations until the water supply at the scene ran low and then he took a break.

Lieutenant Westover complained to other firefighters of a headache and went to rest in the cab of an onscene apparatus. Onscene ambulance responders found him in the cab exhibiting signs of a heart attack and began treatment. Lieutenant Westover was transported to the hospital and provided with treatment in the emergency room. He did not survive the heart attack.

The structure fire was incendiary.

For additional information regarding this incident, please refer to NIOSH Fire Fighter Fatality Investigation and Prevention Program Report F2009-29 (http://www.cdc.gov/niosh/fire/reports/face200929.html).

October 27, 2009–Time Unknown
Phillip A. Whitney, Fire Chief
Age 72, Paid-on-Call • *Springville Fire Department, Utah*

Chief Whitney responded to the scene of a fire alarm activation. He determined that the alarm was due to the use of a smoke making machine by the building occupants, reset the alarm, and closed the incident at 1821 hours.

The next morning, Chief Whitney called the fire station to say that he was not feeling well and that his arrival at the station would be delayed. At approximately 1015 hours, Chief Whitney was discovered unconscious in his home. He was pronounced dead in his home by arriving EMS responders. The cause of death was listed as a heart attack.

November 3, 2009–1655 hrs
Robert Paul "Stoney" Stone, Jr., Firefighter
Age 48, Volunteer • *Amity Fire and Rescue, Pennsylvania*

Firefighter Stone returned to his fire station after responding to a residential carbon monoxide alarm. Firefighter Stone began complaining of chest pains and was subsequently transported to the hospital. After undergoing surgery for an aortic separation at the hospital, Firefighter Stone went into cardiac arrest and was pronounced dead at approximately 0730 hours the following morning.

November 4, 2009–Time Unknown
Chad Eric Greene, Deputy Chief
Age 34, Career • *Union Cross Fire and Rescue of Forsyth County, Inc., North Carolina*

Chief Greene completed his shift at approximately 0830 hours due to a late call for a motor vehicle crash. Chief Greene left the station, dropped his son off for preschool, and then went home. His wife returned home and found him unresponsive, called 9-1-1, and began CPR. He was transported to the hospital where he was pronounced dead at approximately 1230 hours.

November 21, 2009–1554 hrs
Walter G. Hessling, Firefighter
Age 54, Volunteer • Dix Hills Volunteer Fire Department, New York

Firefighter Hessling responded to motor vehicle crash at 1033 hours. He suffered a stroke at 1554 hours. Firefighter Hessling died as a result of his illness on November 27, 2009.

November 22, 2009–1200 hrs
Terrance D. Freeman, Sr., Firefighter
Age 37, Career • Rockford Fire Department, Illinois

Firefighter Freeman completed a shift in which he responded to several emergencies at 0800 hours. He suffered a heart attack at 1200 hours and did not survive.

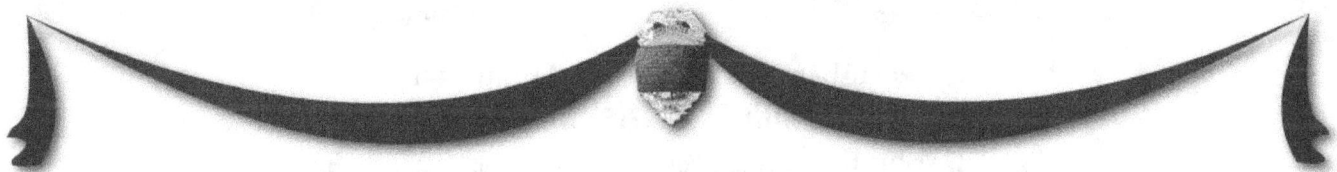

December 4, 2009–1803 hrs
Gary F. Neidig, Jr., Fire Police Officer
Age 36, Volunteer • Mount Carmel Volunteer Fire Department, Pennsylvania

Fire Police Officer Neidig worked traffic control at the scene of a mutual-aid vehicle crash with entrapment. He cleared the scene and went home. Within 2 hours of leaving the scene of the incident, Fire Police Officer Neidig died as the result of illness related to asthma.

December 10, 2009–0655 hrs
Jimmy Lee Davis, Fire Chief
Age 63, Volunteer • White Oak Volunteer Fire Department, North Carolina

Chief Davis was preparing to leave his residence to respond to an alarm for a motor vehicle crash. Chief Davis suffered a medical emergency and collapsed. Although Chief Davis was subsequently treated by first responders and transported to the hospital, he succumbed to his injury. His death was caused by a heart attack.

December 13, 2009–1000 hrs
Urban Aloyisous Eck, Fire Captain
Age 51, Career • *Wichita Fire Department, Kansas*

Captain Eck assisted with fire department operations at a second-alarm apartment fire. He and his crew operated at the scene for over 4 hours. While in rehab at the scene, Captain Eck's vital signs were elevated. Captain Eck and his engine company returned to service and responded to two additional emergency incidents prior to going off duty at 0800 hours on December 14, 2009.

Captain Eck continued to have physical symptoms. He reported for duty his next shift on December 16, 2009, but reported being tired and congested. Captain Eck was admitted to the hospital on December 18 and was diagnosed with heart failure.

Heart surgery was performed on December 29, 2009. Captain Eck did not recover from the surgery and died on January 2, 2010. His surgery was an attempt to repair damage to a heart valve, an acute event caused by the extreme physical exertion at the December 13, 2009, fire.

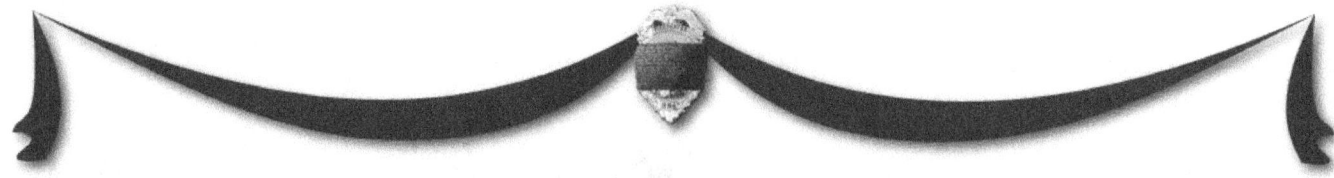

December 20, 2009–Time Unknown
Bobby Joe Mullins, Assistant Fire Chief
Age 52, Volunteer • *Dante Volunteer Fire Department, Virginia*

Assistant Chief Mullins began to experience chest pains while operating at a vehicle fire. He was treated and transported to the emergency room. Chief Mullins was then transferred to a trauma center but passed away soon after arriving of sudden cardiac arrest. His death was caused by a heart attack.

December 24, 2009–1911 hrs
Craig C. Starr, Fire Chief
Age 44, Volunteer • *Plymouth Fire Department, Utah*

Chief Starr suffered a heart attack while transporting a fall patient to the hospital. He was transported to the hospital but did not survive.

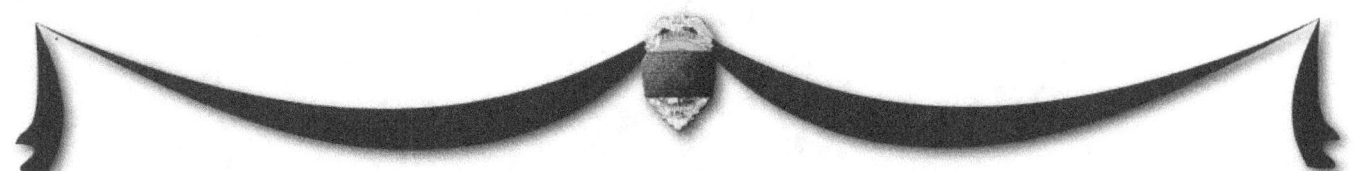

December 26, 2009–2321 hrs
Paul D. Holmes, Firefighter Paramedic
Age 37, Career • Douglas County Fire Department, Georgia

Firefighter Paramedic Holmes was the passenger in an ambulance responding to a vehicle fire with injuries. As the ambulance attempted to pass another car traveling on the same direction as the ambulance, the car struck the ambulance and caused the driver of the ambulance to lose control.

The ambulance fell onto its side, slid, and flipped 3 times. The ambulance came to rest on its wheels. Firefighter Paramedic Holmes was ejected during the crash. He was transported to the hospital where he died as the result of traumatic injuries on December 28, 2009.

December 26, 2009–Time Unknown
Clair Melvin Pierce, Firefighter
Age 68, Volunteer • Wellsboro Fire Department, Pennsylvania

Firefighter Pierce passed away at home after responding to a series of EMS and fire calls. The cause and nature of Firefighter Pierce's fatal injury are unknown.

December 29, 2009–1920 hrs
Steve Koeser, Firefighter
Age 33, Volunteer • Saint Anna Fire Department, Wisconsin

Firefighter Koeser and the members of his fire department were dispatched to a dumpster fire that had been reported by a law enforcement officer on patrol. The incident occurred at a local manufacturing company.

As firefighters applied water to the burning dumpster, an explosion occurred. Firefighter Koeser received fatal injuries and eight other firefighters were injured. An investigation concluded that the explosion was caused by the introduction of water and foam into the burning dumpster. The cause of the fire in the dumpster was not known.

December 29, 2009–1736 hrs
Richard Adam Miller, Firefighter
Age 24, Career • *Belmont Fire Department, North Carolina*

Firefighter Miller suffered a heart attack while engaged in physical fitness activities in the fire station. Firefighters found him unconscious near a treadmill. Firefighters provided medical treatment, including the use of an AED, and Firefighter Miller was transported to the hospital. He did not recover and was pronounced dead at the hospital.

FIREFIGHTER FATALITIES FROM PREVIOUS YEARS

November 24, 1996–Time Unknown
Ramon E. Hain, Firefighter
Age 37, Career • *Saint Paul Fire Department, Minnesota*

Firefighter Hain was exposed to bodily fluids during an EMS incident in 1996. He developed a heart condition that was related to his exposure. He retired from the fire department in 2000. He died on November 14, 2009. He was 50 years of age at the time of his death.

October 20, 2008–1002 hrs
Robert Johnson, Firefighter
Age 75, Volunteer • *Mahopac Falls Volunteer Fire Department, New York*

Firefighter Johnson was assisting with a fire safety demonstration at a nursery school. During the demonstration, he fell from the jump seat area of a fire truck at the school and sustained serious head injuries. Firefighter Johnson remained in a coma until his death on July 5, 2009.

Firefighter Fatality Inclusion Criteria—National Fire Service Organizations

The National Fire Protection Association (NFPA), the National Fallen Firefighters Foundation (NFFF), the U.S. Fire Administration (USFA), and other organizations individually collect information on firefighter fatalities in the United States. Each organization uses a slightly different set of inclusion criteria that are based at least in part on the purposes of the information collection for each organization and data consistency.

As a result of these differing inclusion criteria, statistics about firefighter fatalities may be provided by each organization that do not coincide with one another. This section will explain the inclusion criteria for each organization and provide information about these differences.

The USFA includes firefighters in this report that died while on duty, became ill while on duty and later died, and firefighters that died within 24-hours of an emergency response or training regardless of whether the firefighter complained of illness while on duty or not. The USFA counts firefighter deaths that occur in the 50 States, the District of Columbia, and U.S. protectorates such as Puerto Rico and Guam. Detailed inclusion criteria for this report appear starting on page 2 of this report.

For 2009, the USFA reported 90 onduty firefighter fatalities.

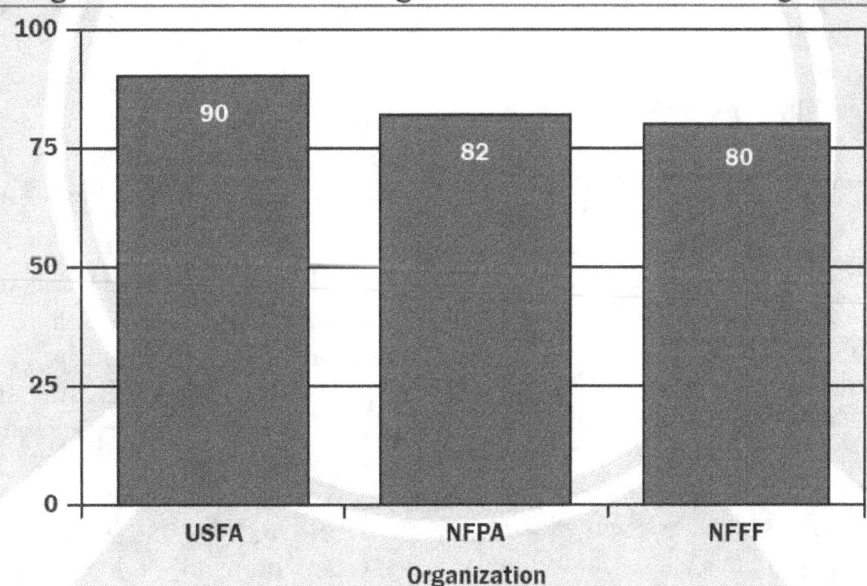

Firefighter Fatalities Resulting from Incidents Occurring in 2009

INCLUSION CRITERIA FOR THE NATIONAL FIRE PROTECTION ASSOCIATION'S ANNUAL FIREFIGHTER FATALITY STUDY

Introduction

Each year, NFPA collects data on all firefighter fatalities in the United States that resulted from injuries or illnesses that occurred while the victims were on duty. The purpose of the study is to analyze trends in the types of illnesses and injuries resulting in death that occur while firefighters are on the job. This annual census of firefighter fatalities in its current format dates back to 1977. (Between 1974 and 1976, NFPA published a study of onduty firefighter fatalities that was not as comprehensive.)

What is a Firefighter?

For the purpose of the NFPA study, the term "firefighter" covers all uniformed members of organized fire departments, whether career, volunteer, a combination, or contract; full-time public service officers acting as firefighters; State and Federal government fire service personnel; temporary fire suppression personnel operating under official auspices of one of the above; and privately-employed firefighters including trained members of industrial or institutional fire brigades, whether full or part time.

Under this definition, the study includes, besides uniformed members of local career and volunteer fire departments, those seasonal and full-time employees of State and Federal agencies who have fire suppression responsibilities as part of their job description, prison inmates serving on firefighting crews, military personnel performing assigned fire suppression activities, civilian firefighters working at military installations, and members of industrial fire brigades. Impressed civilians would also be included if called on by the officer in charge of the incident to carry out specific duties. The NFPA study includes fatalities that occur in the 50 States and the District of Columbia.

What Does "On Duty" Mean?

The term "on duty" refers to being at the scene of an alarm, whether a fire or nonfire incident; being en route while responding to or returning from an alarm; performing other assigned duties such as training, maintenance, public education, inspection, investigations, court testimony and fundraising; and being on call, under orders or on standby duty other than at home or at the individual's place of business. Fatalities that occur at a firefighter's home may be counted if the actions of the firefighter at the time of injury involved firefighting or rescue.

Onduty fatalities include any injury sustained in the line of duty that proves fatal, any illness that was incurred as a result of actions while on duty that proves fatal, and fatal mishaps involving nonemergency occupational hazards that occur while on duty. The types of injuries included in the first category are mainly those that occur at an incident scene, in training, or in accidents while responding to or returning from alarms. Illnesses (including heart attacks) are included when the exposure or onset of symptoms are tied to a specific incident of onduty activity. Those symptoms must have been in evidence while the victim was on duty for the fatality to be included in the study.

Fatal injuries and illnesses are included even in cases where death is considerably delayed. When the onset of the condition and the death occur in different years, the incident is counted in the year of the condition's onset. Medical documentation specifically tying the death to the specific injury is required for inclusion of these cases in the study.

Categories Not Included in the Study

The NFPA study does not include members of fire department auxiliaries; nonuniformed employees of fire departments; emergency medical technicians (EMTs) who are not also firefighters; chaplains; or civilian dispatchers. The study also does not include suicides as onduty fatalities even when the suicide occurs on fire department property.

The NFPA recognizes that a comprehensive study of firefighter onduty fatalities would include chronic illnesses (such as cardiovascular disease and certain cancers) that prove fatal and that arose from occupational factors. In practice, there is as yet no mechanism for identifying onduty fatalities that are due to illnesses that develop over long periods of time. This creates an incomplete picture when comparing occupational illnesses to other factors as causes of firefighter deaths. This is recognized as a gap the size of which cannot be identified at this time because of the limitations in tracking the exposure of firefighters to toxic environments and substances and the potential long-term effects of such exposures.

2009 Experience

In 2009, a total of 80 onduty firefighter deaths occurred in the United States, according to the NFPA inclusion criteria.

National Fallen Firefighters Foundation

In 1997, fire service leaders formulated new criteria to determine eligibility for inclusion on the National Fallen Firefighter Memorial. Line-of-duty deaths shall be determined by the following standards:

1. Guidelines and illnesses.

    a. Deaths of firefighters meeting the Department of Justice's (DOJ's) Public Safety Officers' Benefits (PSOB) program guidelines, and those cases that appear to meet these guidelines whether or not PSOB staff has adjudicated the specific case prior to the annual National Fallen Firefighters Memorial Service.

    b. Deaths of firefighters from injuries, heart attacks, or illnesses documented to show a direct link to a specific emergency incident or department-mandated training activity.

2. While PSOB guidelines cover only public safety officers, the Foundation's criteria also include contract firefighters and firefighters employed by a private company, such as those in an industrial brigade, provided that the deaths meet the standards listed above.

3. Some specific cases will be excluded from consideration, such as deaths attributable to suicide, alcohol or substance abuse, or other gross abuses as specified in the PSOB guidelines.

The National Fallen Firefighters Memorial was built in 1981 in Emmitsburg, MD. The names listed there begin with those firefighters who died in the line of duty that year. The U.S. Congress created the National Fallen Firefighters Foundation (NFFF) to lead a nationwide effort to remember America's fallen firefighters. Since 1992, the tax-exempt, nonprofit Foundation has developed and expanded programs to honor our fallen fire heroes and assist their families and coworkers by providing them with resources to rebuild their lives. Since 1997, the Foundation has managed the National Memorial Service held each October to honor the firefighters who died in the line of duty the previous year.

At the October 2010 Memorial Weekend, the Foundation will be honoring 105 firefighters who died in the line of duty. Of those 105 being honored, 80 died in 2009 as the result of incidents that occurred in 2009, and 25 others died in previous years as the result of incidents that occurred in previous years.

A number of U.S. Air Force and U.S. Marine Corps firefighter fatalities are being added to the Emmitsburg Memorial in 2010. These deaths occurred in previous years and are being added to the Memorial. The following section is a listing of the firefighters that will be honored by the Foundation in October of 2010.

Firefighter deaths that occurred in 2009 as the result of an incident that occurred in 2009:

Alabama

Michael Martin Gilbreath
Double Springs Fire Department

Arizona

Eric A. Tinkham
Queen Creek Fire Department

California

Tedmund D. Hall
Los Angeles County Fire Department

Thomas D. Marovich, Jr.
USDA Forest Service, Modoc National Forest

Arnaldo Quinones
Los Angeles County Fire Department

Continued on next page.

Colorado

Thomas L. Risk
Neptune Aviation Services, Inc.,
USDA Forest Service
Contractor

Brett M. Stearns
U.S. Department of Interior, Bureau of
Land Management
Little Snake Field Office

Connecticut

Charles D. Myshrall
North Coventry Volunteer Fire Department, Inc.

Florida

Richard L. Rhea
Crawfordville Volunteer Fire Department

Robert L. Strang
Melbourne Fire Department

Georgia

Paul D. Holmes, Jr.
Douglas County Fire Department

Derek E. North
Lanier County – Stockton Fire Department

Gary D. Street
East Lake Sinclair, Hancock County Station #2

Illinois

Terrance D. Freeman, Sr.
Rockford Fire Department

John W. Jeffers
Wellington-Greer Fire Protection District

Indiana

Jimmie L. Zeeks
Marion Township Rural Fire Department

Kansas

Lyle C. Lewis
Osborne County Rural Fire District #3

Dennis M. Simmons
Stafford County Fire/EMS Department

Kentucky

Terry L. Sharon
Monterey Fire Department

Louisiana

Richard R. Christiana
David Crockett Volunteer Fire Co. #2

Joseph T. Grace
Saint Tammany Fire Protection District #4,
Mandeville Fire Department

Alan M. Hermel
Bossier Parish Fire District 1

Frankie P. Nelson
Shreveport Fire Department

Maine

Debra A. Cole
South Portland Fire Department

Maryland

Charles F. Clough, Jr.
Sudlersville Fire Company, Inc.

Massachusetts

Kevin M. Kelley
Boston Fire Department

Paul J. Roberts
Beverly Fire Department

Mississippi

John D. Thurman
Clinton Fire Department

George A. Wimberly
Stonewall Volunteer Fire Department

Missouri

David V. Grass, Jr.
Ste. Genevieve Fire Department

Dale E. Haddix
Schell City Volunteer Fire Department

William R. Vorwark
Odessa Fire and Rescue Protection District

Montana

Brian J. Buss
Neptune Aviation Services, Inc.,
USDA Forest Service
Contractor

David M. Jamsa
Minuteman Aerial Application,
Bureau of Land Management
Contractor

Continued on next page.

New Jersey

Manuel Rivera, Sr.
Trenton Fire Department

Gary V. Stephens
Elizabeth Fire Department

New Mexico

Michael W. Flynn
Neptune Aviation Services, Inc.,
USDA Forest Service
Contractor

New York

Jonathan S. Croom
Buffalo Fire Department

Mark B. Davis
Cape Vincent Volunteer Fire Department

Walter G. Hessling
Dix Hills Volunteer Fire Department

Richard W. Holst
Huntington Manor Fire Department

Patrick S. Joyce, Jr.
Yonkers Fire Department

Richard J. Layton
Freeport Fire Department

Charles W. McCarthy, Jr.
Buffalo Fire Department

Paul V. Warhola
Fire Department City of New York

North Carolina

Gregory C. Cooke
Salem Volunteer Fire Department

Jimmy L. Davis, Sr.
White Oak Volunteer Fire Department

Chad E. Greene
Union Cross Fire and Rescue of
Forsyth County, Inc.

William G. Parsons
Millers Creek Volunteer Fire Department

Matthew D. Tramel
Town of Pembroke Fire Department

Ohio

Michael J. Darrington
Toledo Fire and Rescue Department

Conrad A. Mansfield
Delaware Township Volunteer Fire Department

Harold M. Sparks
U.S. Air Force, Wright-Patterson Air Force Base

Oklahoma

John W. Adams
Silver City Volunteer Fire Department

Christopher A. Dill
Oklahoma City Fire Department

John C. Myers
Union Chapel Volunteer Fire Department

Nolan R. Schmidt
Hydro Volunteer Fire Department

Pennsylvania

Albert G. Eberle, Jr.
Roslyn Fire Department

Gary F. Neidig, Jr.
Mount Carmel Volunteer Fire Department

Robert P. Stone, Jr.
Amity Fire and Rescue

William V. Thompson, Sr.
Dushore Fire Company #1

Roy E. Westover, Jr.
Westover Area Volunteer Fire Company

Ryan M. Wingard
Strattanville Volunteer Fire Company, No. 1

South Carolina

James E. Cameron
South Chester Fire Department

Texas

Cory J. Galloway
Kilgore Fire Department

James A. Harlow, Sr.
Houston Fire Department

Damion Jon Hobbs
Houston Fire Department

Continued on next page.

Cohnway M. Johnson
Houston Fire Department

Kyle W. Perkins
Kilgore Fire Department

Utah

Phillip A. Whitney
Springville Fire Department

Vermont

Kenneth E. Frizzell, Jr.
Charleston Volunteer Fire Department

John P. Horton
Marlboro Volunteer Fire Company, Inc.

Virginia

Bobby Joe Mullins
Dante Volunteer Fire Department

Jeffrey H. Reed
Pulaski Fire Department

West Virginia

Johnnie H. Hammons
Craigsville-Beaver-Cottle Volunteer
Fire Department

Timothy A. Nicholas
Craigsville-Beaver-Cottle Volunteer
Fire Department

Wisconsin

Steven J. Koeser
St. Anna Fire Department

Dean W. Mathison
Clayton-Winchester Fire Department

Heath J. Van Handel
Wisconsin Department of Natural Resources

John D. Weber
Township Fire Department

DEATHS FROM PREVIOUS YEARS

Arizona

Joseph Juliano
U.S. Air Force

Arkansas

Monex Thomas
U.S. Air Force

California

Patrick G. Henry
CAL-FIRE, Mendocino Unit

Matthew P. Moore
City of Murrieta Fire Department

Commonwealth of the Northern Mariana Islands

Ernie T. Dela Cruz
Department of Public Safety, Fire Division

Connecticut

Donald A. Trotochaud
U.S. Air Force, Laughlin Air Force Base

Florida

Michael B. Douthitt
Broward County Sheriff's Office Department of
Fire/Rescue

Neal Tarkington
Jacksonville Fire & Rescue Department

Illinois

Kerry R. Sheridan
Troy Fire Protection District

Louisiana

Ralph P. Arabie
David Crocket Steam Fire Co. #1

Massachusetts

Vincent P. Russell
Boston Fire Department

Minnesota

Barry K. Delude
Minneapolis Fire Department

Nevada

William Hartsell
U.S. Air Force

New Mexico

Michael J. Hays
Brazos Canyon Volunteer Fire Department

Continued on next page.

New York

Francis T. Keane
Fire Department City of New York

Ohio

Paul W. Swander
Ohio City Fire Department

Daniel H. Yaklin
U.S. Marine Corps, Marine Corps Air Station,
Yuma, AZ

Tennessee

Mansell L. Hopper
Tennessee Department of Agriculture,
Division of Forestry

Robert M. Weber
U.S. Marine Corps, Marine Corps Air Station,
Beaufort, SC

Texas

Luis Osteen
U.S. Air Force

Ray Rangel
U.S. Air Force, Dyess Air Force Base

Virginia

Roy D. Smith III
McGaheysville Volunteer Fire Company

Washington

Eric R. Lyons
Kennewick Fire Department

Wisconsin

Thomas O'Flahrity
U.S. Air Force

Wyoming

Charles R. Kuhns
U.S. Air Force, F. E. Warren Air Force Base